AF352616

LENS,
LABORATORY,
LANDSCAPE

LENS, LABORATORY, LANDSCAPE

OBSERVING MODERN SPAIN

CLAUDIA SCHAEFER

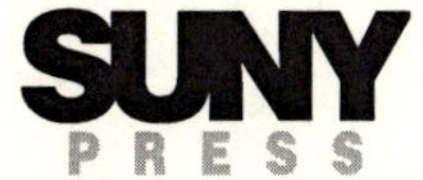

Published by State University of New York Press, Albany

For information, contact State University of New York Press, Albany, NY
www.sunypress.edu

Production by Ryan Morris
Marketing by Michael Campochiaro

Library of Congress Cataloging-in-Publication Data

Schaefer, Claudia, 1949–
    Lens, laboratory, landscape : observing modern Spain / Claudia Schaefer.
        pages cm. — (SUNY series in Latin American and Iberian thought and culture)
    Includes bibliographical references and index.
    ISBN 978-1-4384-5273-9 (hardcover : alk. paper)
    1. Material culture—Spain—History—19th century. 2. Material culture—Spain—History—20th century. 3. Spain—Intellectual life—History—19th century. 4. Spain—Intellectual life—History—20th century. 5. Art and society—Spain—History—19th century. 6. Art and society—Spain—History—20th century. 7. Science—Social aspects—Spain—History—19th century. 8. Science—Social aspects—Spain—History—20th century. 9. Visual perception. 10. Observation (Scientific method) 11. Ramón y Cajal, Santiago, 1852–1934. I. Title.

    GN585.S7S45 2014
    306.0946'09034—dc23   2013036275

10 9 8 7 6 5 4 3 2 1

Contents

Illustrations

Acknowledgments

This book has been both an intellectual labor of love and a personal challenge to find a language of intersecting interests between the humanities and the sciences. In it, I take up a thread of inquiry that began in 1966, and came to fruition some four decades later with the intellectual and material support of a number of very generous sources. In 1966, the unique opportunity to hear Roman Vishniac speak at a conference on microbiology planted the seeds of curiosity as to just how he envisioned that photo microscopy and photography might unite to bring the arts and sciences into fruitful and productive conversation. His slides filled with dark-stained cells and curious microorganisms were matched in intensity by what I later saw in his photographs of the disappearing vestiges of Jewish culture in Eastern Europe. It would take me four decades to come full circle and bring Spanish culture under a similar lens. To Dr. Vishniac I owe both the encouragement of youthful enthusiasm and the seeds of future projects I could not even have dreamed of then.

Lens, Laboratory, Landscape is the product of the most fortunate set of circumstances that an academic might wish for. A University of Rochester Provost's Bridging Fellowship for fall semester 2010 allowed me to consult with colleagues Alyssa Ney and Brad Weslake of the university's Department of Philosophy. Alyssa was both very generous with her time and very patient. I am very grateful for the opportunity to listen in on the discussions of my colleagues in the physics and philosophy group at the University of Rochester. I am equally indebted to Brad Weslake and H. Allen Orr for their generous permission to sit in on their course on Darwin and Religion. Brad continues to be a wonderful collegial resource, and our University of

Rochester Humanities Project on Observation, an outgrowth of class discussions and of this project, has proved provocative and compelling. Without Brad and Alyssa's generosity, I might still be trying to figure out ways of approaching scientific developments in Spain without lapsing into prosodic language or inane suggestions. I truly hope that I have avoided both.

I am extremely grateful to then-Provost of the University of Rochester Ralph Kuncl and to current Provost Peter Lennie and Dean Joanna Olmsted for granting me an additional leave during spring semester 2011. I especially appreciate their understanding of the time it takes to get things right in a new field of research.

This book would not have a solid foundation in the work of Cajal, neither would I have the rights to use his photographs, were it not for Dr. Juan A. de Carlos Segovia of the Departamento de Neurobiología Molecular, Celular y del Desarrollo of the Instituto Cajal in Madrid. In charge of the Museo Cajal, all of the numerous and intriguing slides, photographs, books, documents, drawings, awards, and other items forming the collection, Dr. de Carlos was exceptionally generous, taking time away from his own work. His expertise is incomparable, and the photographs make the man of science "at home" come alive.

On the financial side, I am grateful, and more than somewhat astonished, that the University of Minnesota's Program for Cultural Cooperation Between Spain's Ministry of Education, Culture, and Sports and United States Universities found the resources to support my project with a grant. Given Spain's obviously rocky economic road, this is just short of a miracle, and I thank these institutions with all my heart. I wish the Instituto Cajal and the Museo Cajal equal success and a long future.

In Rochester, I am indebted to the staff at the George Eastman House International Museum of Photography (GEH), who helped me find information about Cajal's photographic work, and the catalog of an exhibit held in Zaragoza. I also owe special thanks to Lisa Wright of the University of Rochester Digital Humanities Center in Rush Rhees Library for her skill with images.

I am especially appreciative of the tremendous patience and fortitude Raúl Rodríguez-Hernández has had in putting up with my random exclamations and endless anecdotes about microscopes and lenses. Raúl's familiarity with the writings of Walter Benjamin has been invaluable, and I am greatly indebted to him for enduring my questions at the oddest hours. In this book, he will finally see the result of all those mornings I spent at the GEH and all the scribbled notes piled on my desk. I look forward to overturning more Benjaminian stones with him in future projects, and he will undoubtedly appreciate me finally cleaning up my desk.

By virtue of their deep lifelong affection and kindness, I cannot fail to thank Helen and Cal, María and Luis, who have always been part of everything. They are forever with me: al fin y al cabo no hay ausentes, siempre están todos conmigo.

Introduction

This is a book about competing cultural and historical constructions of the power and uses of vision and observation as exemplified in the scientific laboratories, photographs, artwork, travel writings, urban development, and cultural geography produced in Spain between the second part of the nineteenth century and the mid-twentieth century. Transformations in the scope of the potency of the use of vision and its popularizing as a mode of knowledge, as a vehicle of discovery, as a road to modernity, and as a corollary to scientific professionalization produced a new language and new values made evident in a variety of material products and popular public images. These products ranged from photographic cameras and development processes to travel diaries, from the microscope slides and biological or chemical findings of the laboratory scientist to the panoply of inventions on display in shop windows, and from artists' reconsideration of the optics of objectivity to cartographers' grids as the visible accompaniments of detailed observation. From an inherited world whose mysteries were approached through conjecture and abstraction, the nineteenth century turned to the eye as the organ of observation and knowledge.

In the discovery of a world of phenomena whose classification and features could be relativized depending on the positioning of the spectator, the choice of subject and the spontaneity of the encounter contributed to systems of social practice in which material artifacts became the weapons of choice in "a battlefield of representations" (Clark 6). Objects such as slides or photographs embodied the results of cautious and insistent observation, but they also reflected competing varieties of experience and training. Observation, both coupled with experimentation in science and uncoupled from it in humanistic practices, brought about convergences and divergences of activities "that straddled the boundary between art and science, high and low science, elite and popular practices" (Daston and Lunbeck 7). Observation drove discovery at home and in the Spanish empire; it also filled notebooks, diaries, navigational charts, travel reports, archives, medical texts, slide collections, photographic albums, and other records of family life. Looking at the world, minutely observing the seen and seeking access to the unseen, observation "is suggestively ambiguous: at once a process, a product, an all-consuming pursuit" (Daston and Lunbeck 7–8). Scientists needed elaborate equipment to do their work; the observer could range from the trained professional to the amateur devotee. Couched in ideologies that shifted across epochs, observation was a powerful tool that taught to seek knowledge in ever-changing places, and to deploy that knowledge in constantly changing ways, including the marketplace.

The curious and challenging relations between the observing subject cast into center focus and privileged, new modes of representation and technological inventions even "persist and coexist uneasily alongside familiar modes of seeing" (Crary 2). The horse and buggy survived the advent of the speedier railway; the illuminating power of electricity did not do away with candles immediately. One might consider the juxtaposition of the traditional family portrait with a collection of stereoscopic laboratory slides. Or think of a weekend stroll on a scientist's gastronomical adventures compared with geographer Manuel de Terán's reasoned cultural observations from a *"geografía de andar y ver"* ["geography of personal contact"] (Anes y Alvarez de Castrillón 10) or philosopher José Ortega y Gasset's curiously coincidental "Notas de andar y ver: viajes, gentes y paisajes" ["Notes on walking and observing: travel, peoples, and landscapes"]. Documents on personal impressions of the countryside, descriptions of startling epiphanies—what Spanish philosopher Paulino Garagorri calls the essentially meandering "diálogo entre el cuerpo y su contorno" ["dialogue between the body and its surroundings"] (10)—and detailed revelations under the lens of the microscope all tend toward the language of scientific experience. Yet, as Daston and Lunbeck clearly point out in their fascinating study, the nineteenth century created a new hierarchy of experimentation (the active intercourse of humans and their environment) over so-called passive observation (3), even as the fruits of observation contributed changes to social life and to the economy. New scientific ideas arose concomitant with a modern society; observation could be a part of the daily life of all, whether in the countryside or in the city. From the elevation of scientist Santiago Ramón y Cajal to international superstar through Spain's fleeting romance with the figure of Albert Einstein, from the material developments of the modern urban environment to the fork in the road represented by Manuel de Terán's precise geographic observations and artist Salvador Dalí's contradictory rejection of the promises of science, observation "cultivates the senses of the connoisseur and straitens the judgment of the savant" (Daston and Lunbeck 7). Observation was an unavoidable, inescapable art as well as science.

The scientific experiments and discoveries of the nineteenth century ushered in a challenge to the mimetic as both an imitative relationship between art and life, and as the complex translation of "the material world [into] a rational order of ideas" (Potolsky 1). A personally inflected recollection of data by the observer was gradually eschewed in favor of experimentation that was extolled as potentially objective and uncontaminated by an individual personality. Yet that very same objective man of science could also be elevated to the status of a hero and celebrity for his discoveries: to wit, Cajal and his Nobel Prize of 1906. Could the professional scientist and amateur observer exist in the same person? Were the scientific results meant to serve other scientists or a general public? Conventions of observation emerged as part of a material culture that embarked on several paths, forming a dialectical relationship between observer and environment.

One arena of social and material development that contributed to bringing techniques as well as objects into the terrain of the observable across social strata was photography. Photography became integrated into a new battlefield of power struggles, and therefore provides markers of those "points of intersection where philosophical, scientific, and aesthetic discourses overlap with mechanical techniques, institutional requirements, and socioeconomic forces" (Crary 8). Photographic images, lenses, laboratories, the camera obscura, ever more portable processes of image development, the expanding toolbox at the disposal of the observer accompanied an increase in mobility via rail lines and roadways. The growing critical relationship between the arts and the sciences are evidence of the value accorded the invention, modification, and perfection of the repertoire of what could be seen, recorded, and deployed. Yet the constellations of products that emerged from observation as the driving force of a whole way of life are simultaneously symptoms of an increased investment in mass visual culture that penetrated the marketplace, the home, the workplace, and the social vocabulary.

That space of urban growth that became the foundation of the modern city was a showcase for the value placed on the eye. Whatever was related to a commitment to optics opened the door to modernity as being "about seeing" (Clark 21). Modernity itself was about seeing and being seen. The flaneur—that "Galicismo intolerable" ["intolerable French import"] (Reglas y consejos 74) to Cajal—was its emblematic, mobile social figure. The exchange value of products and ideas accelerated in a mercantile process, as Jonathan Crary summarizes, "by which capitalism uproots and makes mobile that which is grounded, clears away or obliterates that which impedes circulation, and makes exchangeable what is singular" (10) corresponds to Walter Benjamin's privileging of the flaneur. Goods on display related to the mobility of vision; they could include photographic equipment.

An additional value came to be placed on observation. Crary emphasizes that European modernity, in the end, was related to capital. Observation in the marketplace was cultivated by and for the engines of knowledge and capital that would drive social and economic advancement. New products, new technologies, the dislocation of inhabitants and speed of movement, and the flow of information across telegraph lines, newspapers, radio, and portable photographic images thrust the observer into the quandary of deciphering the barrage of multiple systems of signs. As Lou Charnon-Deutsch explores in her captivating study of the Spanish mass media at the turn of the twentieth century, "the introduction of photomechanical technologies for the direct transfer of information for use in illustrated magazines was transformative" (45). The mass scale of image reproduction used in newspapers and periodicals both responded to and fortified the middle-class desire for the visual and drove consumption.

The possibilities of observation increased exponentially with the innovation of the machinery of the visual. Terán witnessed the migration of rural inhabitants into cities, their abandonment of terrains and the subsequent overflowing of urban resources, demanding the need for analysis after initial observation. As Aurora García-Ballesteros proposes, "Geography in Spain developed as a scientific discipline after the civil war" (1) to rebuild a decimated nation. Terán served as a catalyst for implementing the ideas of intellectuals as varied as Henri Bergson, Pío

Baroja, Claudio Sánchez-Albornoz, Francisco Giner de los Ríos, and José Ortega y Gasset (influenced by German phenomenologist Max Scheler) in Spanish cities. Integrating the social and the natural sciences, Terán turned to Ortega's notions of "acculturation and landscape humanization" (García-Ballesteros 10) to underpin his detailed observations. Terán would publish numerous volumes of photographs, charts, and graphs to accompany his wanderings—"*andar y ver*"—around the peninsula. In turn, these images supported and underscored proposals for careful planning to deal with abandoned landscapes and burgeoning new physical geographies of modern cities. The growth in requirements for housing multitudes was reinforced by photography that accentuated the urgency—and the scientific power—of making the geographer's walk a field laboratory.

While court photography evoked the opulence of the monarch and the power of the military, the photograph itself began to embody a new form of power in the nineteenth century. The magical process that produced an image from darkness revealed a control of physical and chemical processes, knowledge of tools and equipment, and a membership in that group of individuals who had entered the fascinating as well as lucrative territory of this new technology. Owning a camera connected values and desires as much as it connected families, towns, and generations who, with the advent of urban development, might no longer share the same geographical space. The mobile observer as field reporter—José Ortega y Gasset's "El Espectador" ["The Spectator/Observer"] comes to mind—or the social scientist, such as Terán, was deployed to record geographies. Future urban planning evolved. Transmitting the knowledge acquired through observation stood at the forefront of the evidence of the modern; it could convince but also question.

Neither photography nor the development of scientific inventions forms the only subjects of this book. Rather, their lingering cultural effects on the population—artists, scientists, and nonprofessionals alike—take a variety of forms, and adjudicate new power and value to observation in everyday life. While critical to shaping the fortunes of the observer, such products pass on to Spanish society "an uprooting of vision from the stable and fixed relations incarnated in the *camera obscura*" (Crary 14) onto the streets and into the landscape. The stasis and stability of the home darkroom or laboratory was exchanged for the mobility of the open road to offer new angles of vision on old truths. Walter Benjamin concluded that the mid-nineteenth century "deified progress" through the new railways, "contributing to the illusion that industrialism on its own was capable of eliminating class divisions" (Buck-Morss 90–91). The elevation of industrialization and its material products to utopian status changed the experience of the historical, as it also substituted speed for lingering social intercourse. Long-standing structures of power split apart as cities absorbed laborers, educators, scientists, and the industries that they built up. Terán's firsthand reports and images from the field documented this metropolitan growth, migrants' adaptation to their work environment, and the optimal reconfiguration of resources for the most beneficial habitat.

The relationship between modern society and industrial modernization was stuttering at best. Optical technologies were deployed in Spain to document how narrow lanes became avenues, squat buildings grew to new heights, concrete could be formed into unexpected shapes, telegraph lines increasingly dotted the landscape, streets were filled with emigrants from the provinces, and electric lights adorned thoroughfares. Intellectuals coalesced to ponder the mysteries of the natural world and advocated for inventions to acquire knowledge about it, and the sciences rose to occupy privileged status among professionals as well as the masses.

Underpinning my study is the groundbreaking work on observation by Lorraine Daston and Jonathan Crary, who both occupy a crucial place at the cutting edge of the history of observation in early modern and modern culture, the first in the domain of science and the second in art. Their volumes *Objectivity* (with Peter Galison), *Histories of Scientific Observation* (with Elizabeth Lunbeck), and *Techniques of the Observer: On Vision and Modernity in the Nineteenth Century* pose the important questions and offer alternative ways of looking at vision and objectivity. These are not just consequential scholarly studies but excellent reading for cross-disciplinary dialogues.

On another front, not Spanish by birth but born in Germany, philosopher, translator, social theorist, and literary critic Walter Benjamin (1892–1940) informs this book for his influential contributions to the study of the politics and aesthetics of European modernity. His writings on the rise of mechanization in the arts and the loss of an aura of the original, on the structural components of the modern city as emblems of progress, on Charles Baudelaire as lyric poet *and* observer of the contradictions of capitalism, and on the power of photography form a critical part of the general discourse on modernity in the first decades of the twentieth century. Fleeing the Nazis, first in 1932 and then in late 1933, Benjamin wrote the short stories of *Ibizan Suite* and polished the personal recollections of *A Berlin Chronicle* during his stay on Ibiza in the Balearic Islands. His observations on bourgeois houses compared to peasant dwellings continued his study of the architecture of modernity that he had begun in 1927, as documented in the fragments of his *Passagen-Werk* or *Arcades Project*. Buck-Morss underscores Benjamin's fundamental contribution to pan-European debates on modernity in the notes for this project, underscoring that "the theory is unique in its approach to modern society, because it takes mass culture seriously not merely as the source of the phantasmagoria of false consciousness, but as the source of collective energy to overcome it" (253). Of course, although not a stranger to Spain, Benjamin is not always writing about it in particular. But his firsthand consideration of European cities offers fertile ground for the observation of city labyrinths on the Iberian Peninsula.

Starting in the nineteenth century, Spanish urban planners eyed the layout of Madrid and projected the need for a new axis to connect developing neighborhoods. Between 1905 and 1929, the Gran Vía was carved out as a center of transit, entertainment, and commerce. The first three decades of the twentieth century saw architectural experimentation and the construction of what would become iconic buildings of a new, modern city core. As Benjamin noted in other European cities, "the new urban-industrial world had become fully re-enchanted" (Buck-Morss 254), producing a dreamlike state of consumption and advertising that touted consumption. The city itself took on the guise of a "dreamworld" (254) through which one could move as if disconnected from all but products and commerce. Observation of the urban world included signs and images; observation of the natural world would often indicate spaces of abandonment or, conversely, the mythification of now-abandoned space as an indicator of modernity itself.

Neither immune nor isolated from European debates until the advent of the civil war in 1936, Terán returned from study in Paris filled with notions of French geographers regarding culture, observation, and the cohabitation of human beings and their urban landscapes. He would literally walk the cities, as did Benjamin, to observe what the cultural and geographical topographies might reveal. Benjamin hoped for a messianic revelation, Terán a secular one.

As Benjamin theorized, modernity is not a "one-way street" but a complex relationship between past and present, routine and innovation, accumulation and accommodation, private and public domains, and "new forms of cultural and artistic opportunities and awareness, a new relationship between art and audience" (Gilloch, *Myth and Metropolis* 52). Scientific and technological products are also evidence of those processes. Optical devices entertained citizens in movie theaters, exposed the masses to flickering images of faraway places on the silver screen, and opened up the invisible world to the critical scrutiny of the scientist. The man sitting in front of a microscope or passing around a stereoscope to guests no longer merely promoted his interesting pastime but an opportunity to experience the new. Peering through the lens could document the structure of nerve cells as histologist Cajal recorded, but it could also reproduce normative requirements for looking. Representing Spain at world conferences on cell biology, Cajal came in contact with other scientists and their experiments. The observer had become, like the material goods of the marketplace, increasingly mobile, whether in the *"andar y ver"* of Ortega and Terán or in the outsider perspective of Salvador Dalí. Not everyone could be everywhere, but everywhere could be brought to the observer.

Crary observes "the loss of touch as a conceptual component of vision meant the unloosening of the eye from the net work of referentiality incarnated in tactility" (19), and modern technologies contributed to that disconnect. One could observe without direct contact; one could pass through without stopping; one could examine a photograph without having been there when it was taken. Representation took the place of immediate experience. Books, periodicals, scientific publications, and albums were collected, and they disseminated a wealth of visual information at the fingertips of a marketplace avidly interested in capturing the heterogeneity of human life. The mobility of the traveler and the rapidity of the railway system allowed for little direct contact with what the eye could perceive. Instead, a momentary observation permitted the movable subject to capture ephemeral images through word or lens, metaphor or image, light or shadow. So the leisurely strolls of Santiago Ramón y Cajal and his scientific colleagues, poised on a riverbank or inside the dining room of a regional *fonda* [inn] were transformed decades later into the brief encounters of José Ortega y Gasset with the small towns of Castilla or Aragón. They allowed for philosophical reflection, for observations on color and light, or architecture or topographic features but not necessarily for a physical link between observer and observed. In *La rebelión de las masas*, Ortega writes of "*el hombre-masa*" or mass-man as an observable phenomenon, "*visible con los ojos de la cara*" ["visible with the eyes on your face"], "*los ojos en pasmo*" ["wide-eyed with astonishment"], and obvious for "*unas pupilas bien abiertas*" ["alert pupils"] (*La rebelión de las masas* 47). But the lingering effects of what visibly surrounds him remain long beyond the initial impact of the image. Things could be seen without contact, buildings appear isolated from their histories, the old crumbled before the eyes of the observer into modernity's key state: "obsolescence" (Gilloch, *Myth and Metropolis* 110). Time was made visible through its creation of ruins, the vestiges that linger of what once was. As Walter Benjamin observed, time was interjected into the equation of vision and viewer as a primary element of disjunction. The value of memories was equivalent to the quality of the image and its capacity to evoke.

This project was originally conceived with a focus on the concept of retinal vision in the essays of Spanish philosopher José Ortega y Gasset. It soon became obvious, however, that it would be an intellectual miscalculation to isolate his essays from previous scientific and social debates. Public discourse in Europe between the 1830s and 1950s could not avoid engaging with "demon sight" (Stuart Clark 25) in both the sciences and the humanities. I therefore made the decision to delve into the work of Cajal as a fundamental forerunner of a European Zeitgeist, the general cultural milieu, during the first decades of the turn of the twentieth century. It turned out to be not just the right intellectual decision, but also the door to a series of fortunate and unexpected discoveries.

This book is built on an integrated interdisciplinary approach to the material culture of a modernizing Spain—including products of the arts, natural sciences, philosophy, and the metaphorical afterimages of quantum physics—and how objects and images were assimilated in the sciences and the arts. I foster a dialogue both among disciplines and across cultures of potential interest to scholars on both sides of the Atlantic. The project, based on evidence harvested from Spanish history and culture enhanced by other European cultures, enhances arguments for the primacy of visuality but also addresses transitional moments and how they provoke "a crisis of assimilation" (Crary 23). Not only are there more things to see and to ingest, the constant incentive both economically and intellectually to keep searching for something new is part and parcel of modernity. Friedrich Nietzsche analyzes the dislocation of the observer when faced with "the abundance of disparate impressions greater than ever: cosmopolitanism in foods, literatures, newspapers, forms, tastes, even landscapes. The tempo of this influx *prestissimo*; the impressions erase each other" (*The Will to Power* 47). No longer contemplative, the observer engages, if but momentarily, with a flood of images wherein the value of the eye equates with the management of what is seen.

Inventions, theories, and practical instruments all reflect the restructuring of the ideas of space and time, of energy and entropy (chaos, disorder, capital consumed), light and reason, understanding and doubt, the natural and the cultural, and the mimetic and the nonmimetic. The evolution of scientific discourse and the evolution of social discourse go hand in glove, in Spain as elsewhere, as the traces of an imperial state are discarded as remnants of the old in favor of the production of a modern state. Modernity is never born whole, but in fits and starts, in the coexistence of old and new technology, and in mass reproducibility. Phillip Prodger summarizes this movement as a conjunction of the viewer's "expectations," the look of scientific "authority," and the "credibility" of the best and most famous figures (xvii). Santiago Ramón y Cajal, José Ortega y Gasset, Salvador Dalí, and Manuel de Terán Alvarez joined the ranks of Charles Darwin and Albert Einstein as proponents of new thinking that would usher in a modern Spain that would be visible to professional audiences and comprehensible to Spaniards across the board. The crucial intersections of intellectual life in Spain with colleagues elsewhere brought Spain into a general shared discourse of science, art, and philosophy. As Egon Schwartz concludes, in the first decade of the twentieth century "Spain prominently reentered the intellectual concert of the Western world, mainly through the efforts of a few gifted poets and essayists . . . [whose] works were speedily translated, avidly read, and heatedly discussed abroad. And they, in turn, regarded the reabsorption of European culture by Spain as the principal requisite for their program of national revitalization" (87). Ortega in particular looked to Germany for such restorative influences; Terán followed the "science of landscapes" (García-Ballesteros 11) of French geographers. For his part, Dalí absorbed the ideas of German psychologist Sigmund Freud as well as the revolutionary aesthetics of the French Surrealists and the scientific experiments of nuclear physicist, catastrophe theorist, and French mathematician Rene Thom.

By weaving together narrative strands of inquiry, artifacts, and cultural systems as evidence I pay attention to the select few things of the natural and social worlds that became the object of sustained focus in Spain (as in the rest of Europe) between the 1830s and 1950s. These include objects related to one another through the fields of photography, geography, cartography, cinematography, journalism, the arts, urban landscapes, and architecture. They focus on use in the home, family, work, and social and political life. I give heightened consideration to the value of the empirical that enhanced what natural vision could only conjecture about. I examine the comprehensive influences of what Gerald Holton (2001) refers to as a Zeitgeist in which modern European thought on science and its material culture arose. Holton examines the philosophical notions of time and spatial representation as illustrated by mathematician-philosopher Henri Poincaré, artist and sculptor Marcel Duchamp, and physicist Albert Einstein. He roots out the creative innovations of both artists and scientists, all owing a debt, directly or indirectly, to the rise in popularity of non-Euclidean geometry in the early part of the twentieth century, and all interested in the role of intuition in the process of scientific discovery.[1]

Not just what to see but how to represent what is seen signals a change in practices of image-making as well as a change in the relationship between the scientist and his or her knowledge of the natural world. Daston and Galison view the virtue of blind sight—so-called scientific objectivity—as the drive to "rewrite and re-image the guides that divide nature into its fundamental objects" (16). Ramón y Cajal stands out as a figure whose work in the laboratory on neurons and in the darkroom on chemical processes permits us to study "the division of nature into its fundamental objects" and how such classification is conceived and reconceived. Perceiving phenomena under the lens of the microscope and their subsequent artistic rendering placed Cajal at the center of innovation, veneration, and obsolescence related to ways of seeing. Using the stain created by Italian pathologist Camillo Golgi, Cajal was a pioneer in studying and drawing brain cells. In a life that spanned many social and technological transitions, Cajal also promoted and practiced photography, from collodion emulsions to portable cameras.

This book considers the philosophical and material influences of a radical shift in "ways of seeing," what Martin Jay terms "the new ontology of sight" (vii), in fields as disparate yet enticingly intersecting as the arts, physics, the natural sciences, and philosophy as practiced in Spain (within a greater European context) between the 1830s and the 1950s. With new objects of inquiry—lenses to study cell structures, engineering devices, architectural advances, movie projectors, single-lens and multiple-lens cameras, color photography—arose new artifacts, new images, new languages, and new heroes. Among the latter, the scientist—in his many and varied guises—held sway over them all. As science emerged from earlier gentlemen's societies into the world of professional disciplines, the figure of the individual acquired greater recognition, increased discipline and, most notably, a public profile of authority and reverence. What would start with the acknowledgment and awareness of dedicated scientists would later become the iconization and politicization of such figures as Darwin, Cajal, and Einstein, ending with what has become the Darwin industry or the marketing of Einsteiniana. The "international community of scholars who have dedicated themselves to studying Charles Darwin's life and work" (Prodger xi) has parallels in the Einstein industry or that corpus of scientific and political figures dedicated to the promotion of Cajal's work. Suffering a less organized fate than the correspondence, papers, memorabilia, and photographs of Darwin or Einstein, the combination of the Instituto Ramón y Cajal and the, for now, partly in storage Museo Ramón y Cajal remains nevertheless an amazing collection of drawings, albums, histological sketches, photographs, memorabilia, books, slides, medals, and awards including his 1906 Nobel Prize, and letters that attest to the diverse interests and growing visibility on the world stage of that distinguished Spanish scientist.

Emergent technologies, scientific discoveries, the foregrounding of the role of the observer, a simultaneous intoxication with the figure of the scientist as superstar, and a plausible trepidation at the possible uses of any investigation, change the questions asked by intellectuals about both the natural world and the cultural world. The sometimes bold, sometimes more cautious, steps into modernity require a new look at evidence, a serious reconsideration of taking conclusions on faith, and a relativizing of process and product. This is true from Einstein's theories on light to Henri Poincaré's maps, from spatial design to cartography and urban planning, from the consequences of Surrealism's reconsideration of tradition to the development of color photography as a rival to nature. This book examines critical discoveries, radical propositions, and innovative work by neuroscientist Santiago Ramón y Cajal; philosopher José Ortega y Gasset; philosopher Walter Benjamin; artist and provocateur Salvador Dalí; educator, humanist, and scientist Manuel de Terán Alvarez; writer and educator Pelayo Vizuete Picón; and theoretical physicist Albert Einstein from the 1839 arrival in Spain of Daguerre's manual on photography to a mid-twentieth-century fascination with stereoscopy (both Cajal and Dalí), and the quantum field theory and particle physics of Werner Heisenberg.

With the invitation extended to Albert Einstein to speak at Madrid's unique cultural center and residential college—the *Residencia de Estudiantes*—in 1923, and as a potential *catedrático* ["university chair"] in the same city in 1933, the figure of the scientist as the bearer of modernity became unshakably rooted as a staple of the popular imagination. This iconization corresponded to Ortega's own desire to bring the modern world into Spanish thought and culture, as well as to the Minister for Education and the Arts Fernando de los Ríos' official plan to convene a body of learned scientists around Einstein and establish a laboratory for them in Spain. In that way, readers of Ortega's essays in the newspaper *El Sol* could share in the popularization of the giants of scientific discovery, even if they were not informed enough to comprehend all the complexities. Men of science were exalted. Writer Ramón Pérez de Ayala and others who belonged to intellectuals of the Second Republic envisioned the founding of a research institute bearing Einstein's name, but the German scientist's declining of the chair short-circuited their plan. Spain was left, however, with the increasingly renowned Cajal.

There were many inroads made by science and scientists into the discourse of Spanish culture. Pelayo Vizuete Picón's 1923 popular booklet proposing to explain to the masses the basics of Einstein's theories of relativity—entitled *Einstein y el misterio de los mundos* [*Einstein and the Mysteries of the Planets*]—was just one example of publications that acted as bridges into the rarified world of physics and astronomy. It was deemed crucial for the nation to participate in scientific discourse in order to be considered part of modern Europe. How Nobel Prize winners Cajal and Einstein fed into the vernacular world of observation, and signified the embodiment of necessary knowledge—beyond the lab and into the darkroom or the living room—offers insight into a complex Spanish culture during the first decades at the turn of the twentieth century.

I begin my study with the microscope used by histologist and artist Santiago Ramón y Cajal to disprove inherited theories of cellular structures and the connections between the eye, neurons, and the brain. His subsequent turn to photographic lenses to reproduce the function of the eye found links among retina—light and graded or color images—landscape, and portraiture. Cajal's advancement of microscopic observation predicted what would become a change in traditional scientific knowledge while his development of photographic processes recorded that shift in the social sphere. While his portraits form repositories or albums of the faces of human beings who have come and gone, material vestiges captured, frozen in time and space until the observer awakens them once again, the microscope similarly captures the nuances of the natural world as a realm of details also subject to the limits of perception over time. The very small, the very distant, and the very large all escape regularized standards of expectation, yet with the aid of the lens, more can be brought into view. The science of the laboratory, in addition to the scientific refinement of the photographic process, furthered the gathering of evidence, the fixing of images caught with better detail, and the collecting of specimens for generations to access in their quest for repetition, scientific explanation, and understanding. The continued scientific use of Cajal's histological slides long after his death, along with the information made available to others through his experiments, are examples of his foundational contributions. His photographic images record the surrounding social context of those experiments and provide an additional legacy.

From Cajal, I move on to the essays of philosopher José Ortega y Gasset composed during his travel across Spain and Europe. Buck-Morss discerns that "railroads were the referent, and progress the sign" (91), as notions of space and time were forever altered. By this I do not just refer to actual travels, but also to how the extraordinary year 1905 for Einstein (when his articles on space, time, and mass were published) opened new ground for physics, perception, and the relativity of observation. Einstein's theories and Ortega's philosophical humanizing of the landscape are brought into focus using Benjamin's studies on photography, commercial arcades, and the ruins of the city. Within the celebrations and crises of modernity, I focus on Spanish culture's interpretations of Einstein, and on the enticement of the new physics for artists and scientists alike—what Gerald Holton so eloquently terms an "interdigitation of one part of culture with another"—(133) as a conduit to the modern.

I bridge the world of science and its reception by the general public through a reading of a popular pamphlet by Pelayo Vizuete (1872–1933), educator and author, that offered an interpretation of Einstein's universe for the masses. I conclude an examination of the legacies of Cajal as a scientific figure by turning to the Catalan artist Salvador Dalí as an example of Surrealism's troubled relationship to such icons. Dalí's deployment of science as a subject for art not only united two disparate means of observation—the impassioned artist and the dispassionate observer—but also simultaneously reenchanted and disenchanted the myth of a singular, utopian modernity. Dalí's use of multiple images, duplicate forms, bodies opened to observation by the naked eye, stereoscopes, and the idea of the material world as irrational illusion marks a step away from objectivity and a leap toward observation as disenchantment. From Cajal to Terán, the science of the eye underwent many changes. Yet it left behind salient questions about the afterlife of the observed image and its evolving value in a world of bellicosity and destruction.

Notwithstanding Spain's difficult political and economic situation in the 1830s that made a climate of scientific progress difficult to imagine, Daguerre's 1839 breakthrough manual on photographic reproduction reached the hands of Spanish entrepreneurs. As a young man, histologist Santiago Ramón y Cajal (1852–1934) inherited the traditional techniques of the laboratory associated with scientific inquiry (microscopes and handcrafted ink drawings), but he also found photography a tool in the arsenal of the scientist's dream of observation without personal intrusion. He translated the requisite values of the lab into images of land, family, and urban landscapes.

The same year Daguerre's discoveries were presented to the public in Paris (1839), three translations of the manual on the daguerreotype reached Spain. Several decades later, fascinated by chemical processes as well as possibly improving the quality of the resulting images, Cajal took on the study and refinement of the photographic image in the laboratory *and* in the darkroom. His translations from one type of representation to another, from microscope to sketch, then from camera to notebook, reflected as much as they triggered the transformation of knowledge from theoretical into material objects. The value of these objects in philosophical terms was as evident to him as the potential economic value (something he approved of much less), both gleaned from the virtues of empirical experimentation and investigation. For Cajal, the human eye became an object of study in the surgery as much as an organ whose workings might conceivably be imitated or intensified by a lens. The silver nitrate solution eventually used to develop detailed and multiple photographic images of his home, wife, and children was the same solution he employed to capture with utmost clarity the invisible world revealed under the lens of the microscope.

A consummate seeker of objectivity in his scientific endeavors, Cajal is the first scientific superstar in Spain, simultaneously a man of letters and a man of science. While many know of his 1906 Nobel Prize for the histological studies that will later become the foundations of modern neuroscience (as the groundbreaking studies by Laura Otis in *Membranes: Metaphors of Invasion in Nineteenth-Century Literature, Science and Politics* and by Dale J. Pratt in *Signs of Science: Literature, Science, and Spanish Modernity Since 1868* indicate), his 1912 manual on comparative color photographic processes remains unstudied critically and contextually, as do his family photographs, still life photographs, and landscape photographs. Ordinarily relegated to the position of addendum to his scientific work, they represent much more than a way to pass the time. They are, as Roland Barthes suggests, links among images, time, and death that Cajal described in the paradoxical terms of the cinematographer's dream to rewind the reel to bring the dead back to life, a fact obviously less than scientifically supported.

Cajal emerges as the just first pillar of this study as he bridged the epistemological shifts between the nineteenth and twentieth century's ways of seeing that trace parallel shifts in social and cultural values in Spain. The first two chapters of this book fix a European intellectual history into which Cajal's work on cells and on photography is inserted. This cultural and scientific context is developed further in chapters 3 and 4 by examining the responses of Spain's intellectuals to the optics of art and science, possible collaborations among fields, the relationship between human agency and noninterventionist objectivity, the cultivation of the scientific self, the shift of the medium of representation from the verbal to the visual, and, of course, the instruction of the audience (society) in the interpretation of the visual. Cajal is not the single, direct cause of all the subsequent debates, but his skill in recording images, his tireless dedication to observation, his intuition about the role of photography in modern art and commerce, and the fate of his own image as superstar make him a splendid point of departure.

Chapter 3 includes an exploration of the abundant metaphors of light and clarity alongside speed as metonymies of the modern in the writings of Cajal, Ortega, and Benjamin. The association of luminescence and human comprehension is followed by the introduction of Ortega y Gasset's commentaries on retinal observation. In chapter 4, I juxtapose the topographies of Spain as an album of visible and rational geographies (Manuel de Terán) with the negative yet creative value of uncertainty for Salvador Dalí, whose library contained vast collections of scientific books and journals. Walter Benjamin found that Surrealists "became stuck in the realm of dreams" (Buck-Morss 261) when it came to confronting emergent consumer culture. For Benjamin, the response had to be an awakening or an opening of the eyes from the dream, a prying of the image from the retina and embedding it into the processes of history. Dalí remains in his dreamworld, engaging with science on his own terms and reaping the financial benefits.

While scientists have engaged in numerous discussions about the domains of thought woven in and around their fields, the discourse of modern science has also impacted, I believe, many other areas of inquiry. It encompasses how human beings ask questions about their surroundings in the natural world, why they ask certain types of questions, who is authorized to posit them and to whom, and what sort of evidence is required to reach conclusions. This opens a variety of doors into the language of inquiry itself, the reliance on so-called fact, and a focus on the hierarchy of the organs of perception as the tools human beings have at their disposal. In some ways at least, the investigatory path—the nature of inquiry and its conflicted outcomes—is infinitely richer in nuance and information than a need for a binding or final conclusion.

Philosophical and scientific investigation in Spain has often thrived, ironically, on critical input from outsiders and exiles. Exile as punishment as well as opportunity points to contact with a broader Europe and a vast American empire, as well as a deeper philosophical and scientific engagement with human life and with other social and cultural worlds. From Cajal to Dalí through José Ortega y Gasset,

Pelayo Vizuete, and Manuel de Terán, dialogues of Spanish intellectuals with their European counterparts and their meditations on procedures, ethics, and goals, opened the door to placing scientific work and image-making in a complex and productive realm of discussion.

This study addresses historical and cultural moments at the turn of the nineteenth century when the intellectual debates of the first few decades of the twentieth century originated. It explores the social spaces impacted by innovative forms of technologies and the material results of experimentation with light, lenses, and the reproduction of images. It seeks how inherited values were challenged, remythified or rejected, or how philosophers, artists, writers, scientists, social scientists, and physicists assimilated them differently into their thinking. Fueled by a radical sense of change across the arts and sciences of the European continent the revolutions in the physical sciences, both proposed a reconsideration of traditional analytical processes and conferred increasing authority on emergent scientific discoveries, observations, and experiments. The impact on art and social sciences of "both non-Euclidean geometries and the populist, idealist, sometimes mystical spaces of higher dimensions" (Parkinson 2) had repercussions on the creativity and methodology of the work of Dalí and others.

It remains a challenge to the twenty-first-century to approach the lasting debates of how philosophical inquiries, scientific experiments, a cult of the observer, artistic visions and innovations, and material products come together. Changes in perception brought about by the new physics, scientific experimentation, theories of relativity, and, ultimately, technological innovations and discoveries, continue to engage us with inherited values and observations.

Chapter One

The Creation of a New Scientific Persona

Santiago Ramón y Cajal and the
Rise of Popular Photography in Spain

The story of the development of modern science in the West, and the rise of scientific models through which the environment could be either comprehended or controlled, contains a variety of interpretive knots. Among the many components of this narrative history, one focus of scholarship has held fast to the importance of "the extraordinarily rich and complex relationship between science and religion in the past" and that "during their history, the natural sciences have been invested with religious meaning, with antireligious implications and, in many contexts, with no religious significance at all. Not only have the boundaries between them shifted with time, but to abstract them from their historical contexts can lead to artificiality as well as anachronism" (John Hedley Brooke 16). While this statement appears to cover the gamut of possible relations between early modern European natural history dedicated to the description and cataloging of phenomena, and natural philosophy that sought out the causes or sources of those phenomena, it confines any intellectual discussion of the scientific revolution in Spain to very limited quarters. For many, it cancels out any possibility of discussion at all.

There are alternative paradigms, however, that encourage debate and discussion to fill in the blanks and look "elsewhere" for clues to the practice of what we might call the observational sciences in a variety of venues. For example, the volume edited by Daniela Bleichmar, Paula De Vos, Kristin Huffine, and Kevin Sheehan successfully argues against "an image of . . . empires built on the quicksand of superstition and greed," concluding with excitement and accuracy that one finds evidence to support the central premise that "'science' was the handmaiden of the Iberian empires" (Cañizares-Esguerra 10). Scientific practice responded to colonial expansion and documentation, to the empiricism of empire, to the "green treasures" of native fauna found even before the gilded ones sought, and to the necessity of creating "mnemonic aids to help distant audiences experience . . . without traveling" (Cañizares-Esguerra 1, 4). As Daniela Bleichmar strikingly encapsulates the two poles of exploration and material production, the empire's undertakings had to be made both "visible and useful." Archives of exacting images were deemed a record of "the colonial machine [that] was a visual apparatus" (309). Maritime charts and navigational technologies fed the enterprises of collecting and cataloging, but they were the successful recipients of imperial investment as well. Among the growing number of scholarly projects related to "things scientific" that have begun to circulate regarding the knowledgeable intellectuals of the Spanish empire on both sides of the Atlantic, Miruna Achim's excellent overview of the critical shift in such studies over the past two decades makes it abundantly clear that earlier scholars' orientation toward viewing the Counter-Reformation and ensuing cultural closure of the nation had turned a blind eye toward a possibility of scientific riches. The reason is, as William B. Ashworth Jr. writes, "we have not been asking ourselves the right questions" (133).

The presupposition of an anachronistic vision of the world from the vantage point of Spanish culture, even well into the nineteenth century, has subsequently made way for the scrutiny of documents, charts, diaries, notebooks, and albums of specimens related to medicine, engineering, mining and metallurgy, and many other spaces where scientific research was collecting data and recording observations. In addition, comprehending and administering the wealth of the natural world over which imperial Spain had taken control was an urgent goal. Achim concentrates on

> specific cases, communities, and contexts, in order to understand why and how science was practiced at different moments and locations in the Spanish-speaking world . . . and more inclusive definitions of scientists and scientific practice, making room for sailors, bureaucrats, travelers, publishers, and merchants, and for activities like collect-

> ing, trading, legislating, and entertaining, . . . making maps, planning cities, collecting and recording data about plants, animals, minerals, climate, and topography. . . ." (107–08)

These are many of the activities embedded in the state's administration, natural resource strategizing, and transatlantic economic development, and the embedding of them within the cultures of exploration, geography, the collection of written records, and many other fields as well as across continents specifies. But they also bring together the many facets of how science was practiced and observed. Scientific facts and artifacts, alongside evidence of how they have been produced and consumed and by whom, literally cover the cultural landscape and leave behind a legacy of material objects as well as documentary evidence. Spain's traditionally seen failures and absences in the area of theoretical sciences such as astronomy or physics had closed the door to anything but handwringing. On the other hand, a consideration of the central role of scientific instruments and processes in Spanish culture is worthy of more attention. These might include Humboldt's early nineteenth-century information collected in "mobile laboratories" (Nieto Olarte 236), accompanied by a pamphlet titled *Paintings of Nature* that synthesized climate, physiognomy, and "an intense collaboration between art and science" (Mattos 143) through Santiago Ramón y Cajal's microscopy and photomicrography, dark rooms, telescopes, photo-phonographs, and stereoscopic cameras. Both on the Iberian Peninsula and in the colonies, scientific instruments were employed to measure, catalog, calculate the value of, and record the holdings of the empire and its inhabitants. Scrutiny of such material objects and the products resulting from their use makes sense in the context of the adoption of the epistemological values of modern empirical science as part of a collective project of modernizing advanced from the middle of the nineteenth century. This push for the modern is especially visible in the wake of the Spanish–American War as the new century dawns. Dale J. Pratt aptly signals the revolution in the fall of 1868 as a breaking point in the historically difficult relationship between the sciences and the arts in Spain, finding that historical time as one that "has evinced an ever-increasing concern with the implications of scientific inquiry" (3). The turn-of-the century war spurred that shift toward inquiry and observation even more insistently, but the groundwork for a focus on science had been laid even before that critical moment.

While Pratt persuasively studies Spanish cultural modernity reflected in the intertwined discourses of scientific and literary texts, other contemporaneous discourses existed as well. The community of individuals concerned with the role of science grew as communication did. R. V. Jones writes that

> The creative individual is, in a sense, complementary to the society in which he lives, rather as a soloist in a concerto. Both the basic ideas of science and the key inventions of mankind have generally been conceived in the minds of individuals, while the effort to gain the data on which the ideas and inventions have been based, and the subsequent effort to turn them to good account, have required the contributions of many besides the inventor and originator of ideas. So the individual and the community are necessary to one another. ("Complementarity as a Way of Life," 323–24)

Scientific communities—within the territory of peninsular Spain as well as across Europe and the Americas—influenced, informed, and challenged one another in the laboratory, the dark room, and the production and consumption of goods and ideas resulting from inventions and technologies.

I propose that those artifacts are all related to a critical appreciation of the preeminence of vision, the eye, the lens, the retina, and the scopic realm of light. The technologies related to sight were employed in the processes of photographic development, in addition to scientists' microscopic work on nerve cells, the spinal cord, the brain, and the retina, and they even came into play in the cinema. (The spectacle of Hollywood provided Ramón Gómez de la Serna with fodder for his allegorical novel *Cinelandia*, the glittering city filled with "illuminated faces" [105], meteoric "blazes of light" [102], and actors with "burnt-out retinas" [103].) The role of luminosity was to allow a pathway to the brain through the orb of sight, as a power to be harnessed in metaphor as well as in the enhanced material products of modernizing societies. While the natural eye may afford access to the details of the world—first its wonders, and then its anomalies—it may also notably be enhanced through the use of lenses that reveal more than unaided sight. When the object under scrutiny is visible, placed before the eye, knowledge of it had been deemed the most authentic and the most doubtlessly accurate. Yet various models overlap across historical eras, challenging both the methods of seeing and how to read what is seen. Subjects, practices, and institutions went through radical changes during the nineteenth century, redefining the position of the observer, relocating the eye in picture making before a now-absent object, and shifting the desire to document experience into new and innovative forms of mass representation. Time and its passage—that inexorable movement toward death—could be seen in photographs, for instance, but the experience of it could not be direct. Cajal's healthy son in early pictures disappears from later ones, but the narrative that recounts his infirmity and death is not made available to the eye in images. Instead, there is a shift toward invisibility as a marker for time to fill in the missing face (or landscape) that has become the victim of the temporal. Terán would find the same true in the absent spaces of geographical landscapes.

William R. Everdell includes Cajal among "the first moderns," at least in part owing to the content of his study of the "atoms of brain" that figure as the components of greater structures of thought. How Cajal sought to study the forms and relationships of nerve cells and their interrelationships showed the "intellectual origins [of the modern] in an often profound rethinking of the whole mind set of the nineteenth century, the world view that gave rise to speed, industry, world markets" (Everdell 9). That worldview juxtaposed the history of science and constantly appearing inventions, the product of the artist on paper and the image produced through a lens, the "convincing-ness" of rapid photography and old reliable printing techniques, the examination of solar flares or eclipses and previous conjectures about the natural world, and the capturing of objects and movement "faster than the naked eye could see" (Prodger xxiii). The decomposition of entities into their component elements, of light into energy and bodies into cells, required the support of emergent technologies that appeared equally as fast. As a theoretical concept, then, modernity has provoked a complex and ongoing debate about time(s) and culture(s), one given a particular tenor by Daniel Frost, who notes a shift in the idea of "culture" as a marker of the modern, and indicates that from "culti-vate" to "culture" there occurs a change in the perception of *paisaje* or landscape that bridges Spain's nineteenth century to the twentieth. Frost cites Raymond Williams on the advent of capitalist economics as a force that brings industry to replace agriculture and towns to replace rural farms, thereby also reframing landscape as urban culture. Frost links his discussion of economics and the land—or changing notion of landscape—to a record of linguistic shifts such as that indicated by Stephanie Sieburth: "In Spain, where industrialization began later and was not as pervasive as elsewhere in Europe, the change in meaning was more gradual and it is not until the final years of the nineteenth century that Spanish dictionaries begin to register a difference" (Frost 16). The lexical register of rural to urban culture accompanied the economics of modernization as well as a shift in point of view.

As a characteristic of that previous century's thought, the metaphor of "smoothness" (Everdell 9) gave way to new divisions and new tempos. Societies began to cast off assumptions about the ease with which the objective observer could watch processes unfold, or the linkage between moments of perception. That is to say, Cajal's focus on the study of the varied elementary parts of the whole as they related to an organism was one of the fundamental signs of modernity's quests and questions. So was his insistence on the power and potential of the photographic image to freeze time. With an increased capacity for magnification and the invention of new technologies, individual components might grow into new collections of fragments and varying articulations. They might also provide the starting point for even greater detail that, when taken as a whole structure, presented an intricacy previously unsuspected. Everdell likens the scientific deconstruction of the totality in order to reconfigure the elements to artist Georges Seurat's method of "dividing optical perceptions into their discrete elements" (64) in his pointillist manner of representing a scene. The distinct dots of color meld if observed from a distance into graduated and subtle mixtures of tone and shade, forms and shadows. If Spanish society—literally as well as metaphorically—"groped about in the dark as it slowly pushed its way into modernity" (Pratt 130), then Cajal's experiments in histology and photography indicate a turn toward the light to identify—to see "rightly"—the whole as a more problematic yet ultimately still comprehensible entity. The emergence of an image from darkness in a photograph, or the illumination of cellular structure under the microscope, held the potential to shed light on the entire world. Cajal provided light on Spain's faltering process toward modernity.

Yet Cajal knew that scientific evidence did not end that process. Had he reached some sort of ultimate conclusions that no longer required examination, he would not have referred to the ongoing

experiments in "artificial somnambulism and phenomena of sugges-
tion" undertaken in his home in the name of "*ciencia positiva*" ["true
science"] (*Recuerdos* II, ch. III), or the parade of cells, people, and
Levantine landscapes that "*desfilaron sucesivamente por el objetivo de mi
Kodak*" ["passed in succession before the lens of my Kodak"] (*Recuerdos*
II, ch. III) across the decades. To answer Pratt's enticing question posed
at the outset of his study, Cajal provides evidence that science *can* be
consumed as "both a praxis and an aesthetic object" (2). That object
is contained in the visual products of the scientific laboratory and
the photographic dark room (slides and photographs), both as process
(praxis, technology, innovation) and as objects (the innate beauty of
the stained cell and its biological function, the details of the natural
world that science brings to light).

Many of the mentioned relationships among disciplines, wavering
between a similarity and distinction of purpose between the religious
and the scientific, concern both shifting cultural and intellectual boundaries, and particular historical events related to Spain. This includes the
encounter with a "'new world" to be cataloged and comprehended;
the Counter-Reformation and implementation of the Inquisition; the
shift from the rule of the Hapsburgs to that of the Bourbons in an
on-again, off-again romance with modernity being embodied in the
instruments of scientific progress; the Spanish-American War and the
failures of medical and military science to save both the troops and
the colonies. The breakdown of the empire in 1898 presented Cajal
and others with evidence of Spain's inability to use all the resources of
modern scientific progress to make life better, earn the respect of other
modern nations, and "*aplicar [la] ciencia a las necesidades de la vida*" ["apply
science to the necessities of life"] (Cajal, *La psicología de los artistas* 113).

He had experienced the results of an inability to harness the power of scientific knowledge for the "necessities of life" firsthand some
twenty-five years earlier, as a victim of the tropical diseases rampant
in the Caribbean during his short time as a volunteer medical doctor
in Cuba. When faced with staggering human losses and illness in the
difficult climate of the islands, to say nothing of the lack of a spirit
like that of Alexander von Humboldt missing from the Spanish culture of modern times, Cajal wrote of Spain's Cuban defeat with both
cultural and scientific regret. He chose his words carefully to depict
the lost promise of scientific endeavors done either carelessly or without full comprehension: "La media ciencia causa la ruina. . . . Hemos

caído ante los Estados Unidos por ignorantes. . . . Eramos tan ignorantes que hasta negábamos su ciencia y su fuerza. Es preciso, pues, regenerarse por el trabajo y el estudio" ["Half-baked science causes ruin. . . . We have fallen to the power of the United States because of our ignorance. . . . We were so ignorant that we even denied their dominance in science and their strength. We must, therefore, be regenerated through work and through study"] (*La psicología de los artistas* 113, 116). There is no doubt that Cajal's conclusions are a result of the downcast historical moment at the end of the 1890s, but they are also colored by his voluntary service as a medical doctor in Cuba beginning in 1874, a hazardous enterprise cloaked in romantic imaginings. Cajal's call for a collective will to stand tall and be "regenerated" through renewed dedication to intellectual pursuits echoed throughout his entire professional career, and he would be the first to proffer his own work ethic and investigative drive for observation as models.

Filled with boredom amid everyday life in the provinces, particularly in Lérida, the young scientist evoked his earlier readings of the novels of Jules Verne and his engagement with other literary adventurers in order to extend them to his own goals. There was a sense of identification between this reader and the imaginative characters that explored the nether regions of the planet in search of the unknown or the unexplained. For a young resident of Spain, these idealistic travels would first imply a visit to the American colonies, the outer reaches of the empire, and the territories of Charles Darwin's studies in the 1830s. As he recounted his emotional response to the imaginings of Verne and the observations of Darwin, Cajal confessed that:

> Me devora la sed insaciable de libertad y de emociones novísimas. Mi ideal es América, y singularmente la América tropical, ¡esa tierra de maravillas, tan celebrada por novelistas y poetas! . . . Orgía suntuosa de formas y colores, la fauna de los trópicos parece imaginada por un artista genial, preocupado en superarse a sí mismo.

> The insatiable thirst for freedom and for experiencing new emotions devours me. My ideal is America, and more particularly tropical America, oh that land of wonders so celebrated by novelists and poets! . . . A sumptuous orgy of forms and colors, the fauna of the tropics seems something

> imagined by a jovial artist, spending his time constantly trying
> to outdo himself. (cited in Laín Entrago and Albarracín 60)

The obvious exoticism of the American colonies reported by explorers and literati alike, their strikingly exciting flora and fauna, and the dream of gazing on the waves of the Caribbean Sea took him on this mission against his father's wishes.

First Cajal ventured to Puerto Rico and, from there, to Cuba. With the deaths of other physicians in Cuba, new ranks of Spanish doctors filled in to try and treat the cases of malaria and tropical diseases that had decimated the population. The romantic trajectory of exploration and the epidemiological realities would intersect at some point. A young scientist with a flair for drawing and a true utopian impulse for the unknown, Cajal set off filled with exuberance. Yet he returned to Spain within a relatively short time, another victim of intestinal disease, tuberculosis, fever, and parasites. His desire for adventure came face to face with the real conditions of the tropics. This encounter between medicine and the natural world left Cajal pondering the shortcomings of a culture choosing to ignore science and those of his own imagination. Both of these forces founder amid the spreading uprisings of the American colonists against Spain that would ultimately bring a new power to the region, the United States. All-out war would not break out in the 1870s, but it was a growing possibility summarily ignored by the center of the empire that turned a blind eye to the discontent of the islands. From the province of Camagüey where he was first assigned, Cajal saw desolation, isolation, and danger all around him. Spanish soldiers and native Cubans battled the ravages of tropical disease, using all the forces of science at their disposal. But these were not sufficient. While Cajal did not mention, and may have been unaware of, the equally catastrophic suffering of the U.S. troops in the Cuban conflict of the 1890s (although one imagines that the medical community would certainly be informed of such crucial statistical data on both sides of the battle), the knowledge potentially provided by science seems to have failed both sides owing to the shortsightedness of governments and their neglect of the scientific discoveries on the bacteriological front.[1]

Rather than concentrate primarily on previous historical times alone or on scientific investigation in a narrower sense, this study turns instead toward late-nineteenth-century and early-twentieth-century Spain as a crucible of scientific activity in which paradigms were inherited and reconsidered. At the turn of the century, the new epistemological practice of empirical observation had become codified and exalted, but also challenged, in the sciences as well as the arts. After the proliferation of the eighteenth century's scientific conventions of truth-to-nature, nineteenth-century European scientists turned increasingly to a more mechanical objectivity insisting on the elimination, to the greatest extent possible, of the willful intervention of the scientist in how natural phenomena looked or how the artist reproduced images of nature. Even as scientists and their work rose in public stature, their interventions in the production of images supposedly retreated. The rise of progress in modern technological devices, such as the camera—and, with it, photomicrography, microphotography, stereoscopy—and an argument in favor of such an invention as a "distinctly scientific medium" (Daston and Galison 130) was accompanied by "different expectations for objectivity" (Strong 63) to live up to. Changes in perception with the deployment of increasingly powerful lenses accompanied and predicted the social transformations occurring between the nineteenth and twentieth centuries, highlighting a rational and comprehensible world now able to be observed as well as investigated with greater particularity. With one eye at the lens of a microscope and the other fixed on a sheet of paper, the scientist and observer had dual perspectives on the natural world, perspectives that required care and caution if they were to be scientifically valued and socially exalted.

In the realm of technological invention, Paul Martineau refers to photography as a technology that "shortened the distance between the eye and the hand" (7), thereby insinuating a realism of the resulting product that appeared to copy nature without any input from the observer. On the one hand, the camera seemed to show without interference, but the subtle arrangements of objects in a still life or the observation of minute cells and structures still exemplified the notion of interpretation. As Laura Otis, referring to the scientist's techniques with the microscope to acquire the greatest definition of detail, and his subsequent drawing of what was observed, astutely summarizes: "Santiago Ramón y Cajal, Spain's Nobel Prize–winning histologist, is known for his vision." She deftly points to both the concern with sight and a forward-looking attitude. What he saw through the lens of the camera was akin to human beings populating a landscape—the composite of elemental parts from the visible to the invisible, "individual cells and human beings represent[ing] the true origin of will, creativity, and regeneration" (Otis 64). He saw the role of the scientist as an exercise of free will, aided by the apparatus of technology, akin to the freedom he sought in the Americas, accomplished by a trained human being with scientific, creative, and culturally beneficial ends. Like colonies of cells, individual scientists were intellectual leaders of a cultural collective, much as patriarchs presided over the family units comprising coetaneous Spanish society. Cajal reproduced as faithfully as possible what he observed under the microscope, down to the arrows that indicated the flow of blood or nerve activity. These images would orient other scientists and produce accurate depictions for further scientific experimentation.

Yet unless and until others could reproduce an encounter with the image in the same manner, the plausibility and validity of the new techniques of observation rested on two things: the scientist's own meticulous records and capability of reproducing an experiment, and the status of the scientist himself. Cajal did the same for his own image as a scientific investigator as he did for the laboratory experiment through the photographic self-portrait of an intellectual constantly at work. Prodger could not state it more clearly: "Photographs assumed a dual role. They illustrated something, but they were also experiments in their own right. They became more than mere pictures—they became data" (xxiii). Both scientific processes—in the laboratory and in the dark room—developed protocols as they simultaneously cast the scientist into the spotlight. Scientific photography and other sorts of photographs occupy the space of the technologies of the eye, the first as signpost to discoveries and the second as identifier of the man who was the discoverer. Darwin was not a photographer but he advanced the science of photography when he chose to incorporate images into his theoretical books as visual illustrations. If Darwin chose with care the type and number of photographs that would prove his point for the 1872 *The Expression of the Emotions in Man and Animals*, using photographs as evidence of scientific hypotheses, Cajal surpassed that activity by the turn of the century with his expertise in developing plates and experiments in chemical processes that produced photographic specimens to prove his theories.[2]

The domestication of nature, or a search for an ultimate cause provided through scientific examination and collection, were conventions that carried over from the eighteenth century, and such a "truth-to-nature" ideal coexisted for a time with the advent of the codes related to the rise of lenses and mechanical objectivity. The aesthetic virtue of harmony ceded only with difficulty to new visions no matter how convincing they might be, yet some aspects may have coincided simultaneously. As Cantor and Brooke conclude of earlier shifts in interpretive analysis related to the sciences, "To lose the music of the spheres was an intolerable deprivation" (174) for Johannes Kepler, whose research was rewarded later with the acceptance of an aesthetic of the ellipse as the new elegance of the planetary orbits. The standard commentary about Cajal's histological preparations of neural circuitries (with the camera lucida as an aid to drawing), cells of the cerebellum, structures of the retina, and Purkinje cells' dendritic tree that he saw as similar to a grape arbor, includes references to their elegance and texture, their attention to detail. In particular, scientists emphasize their "clarity and beauty . . . [that] are even today awe inspiring" (García-López, García Marín, and Freire 15). Not only did he find the "right" way of seeing the cells, in the process he produced a new sort of beauty. So his avid hypothesis of a *neuron doctrine*, opposed to the *reticular theory* promoted by Camillo Golgi, did not lessen the impact of scientific discovery and clarification but enhanced aesthetically what was seen and, finally, explained structures with the exactitude of the lens and eye. The new grace of Cajal's science was its breadth and inclusive vision, its aesthetics of the product and of the process simultaneously. This is noted in the words of Emil Holmgren who nominated Cajal for the Nobel Prize:

> Cajal has not served science by singular corrections of observations by others, or by adding here and there an important observation to our stock of knowledge, but it is he who has built almost the whole framework of our structure of thinking, in which the less fortunately endowed have had to, and will still have to put in their contributions." (cited in Grant 2)

Sheer intellectual drive and curiosity—"irresistible curiosidad" ["irresistible curiosity"] (*Recuerdos* I, ch. XXVI), "tenacidad" ["tenacity"] (Laín Entralgo 10), "testarudez indomable" ["indomitable stubbornness"] (Cajal, *La psicología de los artistas* 19), or "brío inquisitivo" ["inquisitive spirit"] as his brother Pedro saw it (Cajal, *La psicología de los artistas* 41)—drove Cajal to find visual evidence of his hypotheses. It was not enough to theorize. The lens provided scientists a chance to work with the previously invisible, minute details of natural objects and phenomena, counteracting a lack of material evidence or a faulty reliance on speculation. Sight itself was as exquisitely alluring as what was observed.

Everdell points out in metaphorical terms that the inherited task for the taxonomist and collector "was stamp collecting. A good taxonomist had to be humble, as well as extraordinarily thorough and persistent, like Linnaeus. . . . This kind of tireless single-mindedness was very much in the character of Santiago Ramón y Cajal" (101). Yet there had to be more than mere "tenacity" in the shifting of the limits and parameters of one piece of matter or physical structure, and the beginning of the next. There had to be reason and observation. The boundary between the similarities and the differences of two objects was the central focus of any cutting-edge taxonomist, but any method of observing and judging the structural arrangement of properties could always be challenged by technological innovation. Not to be feared, technology was promoted by Cajal as provocative and helpful in the study of all aspects of the world. As Everdell proposes, taxonomy "is more epistemologically challenging than any other science . . . it makes more innocent assumptions . . . What in fact are you seeing when you classify a thing and give it a name? . . . Why are some categories appropriate for bringing things together and others not?" (104). That all life was part of a continuum, not ascribing breaks or distinctions to individual units but an unbroken chain without end, was an assumption that haunted the science of taxonomy until the turn of the twentieth century. Cajal's work in "a small corner of the learned world" (Everdell 106) in the 1880s was quiet, persistent, and in the beginning somewhat invisible, even when he began to make strides in the study of the characteristics and behavior of nerve cells. A portion of that greater framework of human thinking referred to earlier is made evident when Cajal joined Camillo Golgi in connecting the use of silver nitrate—the chemical that also launched the photographic revolution—to the staining of cells in the laboratory. The advance in science would reveal the art of the human nervous system, and open the door to new nomenclatures.

The drive toward accuracy of Cajal's "framework" was like an architectural structure built carefully from step-by-step observation. It offered accessible and ascertainable knowledge, information that may be observed, and even challenged, by others. With the microscope and the camera, new forms of experience emerged that brought the physical world and the social world in increasingly closer contact. It was deemed incumbent on this scientist therefore to look with insistent accuracy at what appeared through the lens, not long for the imaginings of the past before the shift in the value of experience occurs. With this, as shall be discussed later, the almost mythic aura of wonder and distance felt between human beings and the rest of the natural world was punctured by technology whose development presented new opportunities as well as new investigations. In a way, the loss of aura around the natural world corresponds to Walter Benjamin's description of the effects of mechanical reproduction (film and photography) on the auratic uniqueness of the image. With advancements in technology for the mass media, the singularity and moment of originality of an image fades. Daston and Galison propose that, rather than a vision of singular certainty, objectivity in science "preserves the artifact or variations that would have been erased in the name of truth" (17). Deviations and discrepancies opened up to scrutiny theories about nature, especially through the reproduction of images. Like the implementation of mechanical reproduction and the new media of the early twentieth century—cinematography and photography among them—the microscope provided laboratory scientists an opportunity for repeatable experiments. Thus, they could confirm or contradict what others had postulated; the photographic lens afforded a constant invitation to look anew.

Howard Caygill concludes that when Walter Benjamin wrote of the decay of aura brought about by photography and the cinema, "aura is not a property but rather an effect of a particular mode of transmission" (102). This effect leads to objects and observers being brought into closer contact, but in different times and spaces. When nearness—real or simulated—is broken, the relationship of time and space (history) for the observer is also ruptured. If Cajal's insistent work attempted to emancipate science in Spain from its relegation to the edges of culture at the turn of the century, his fame brought its promise into the center of society where it might be put on display as the preeminent motif of the modern. A simultaneous paradox was produced as he was monumentalized and distanced from the public as a spectacular icon of intellectual activity that only exemplary Spaniards could attain. In other words, as the figure of the scientist was made to feel closer than ever through publicity and journalists' reports, that same individual was rent from the social fabric owing to his extraordinary work. The photographic record of Cajal's private life sometimes contradicts the stature of the public man.

Even as technologies of the eye made more dimensions of the world accessible, bringing into focus the normally unknown and the unseen, they also cast out traditional notions of certainty about human life, the notion of time and the privileged observer. They created a distance between the expected and the innovative as well. This accessibility affirmed the need to interact with and interrogate the past and its images, not accept them as they had been inherited. In Cajal's own words, science is the tool of that inquiry: "La ciencia infatigable nos lleva de sorpresa en sorpresa, y cada invención es un placer arrebatado a nuestros abuelos." ["Science persists tirelessly in taking us from one surprise to the next and each new invention is a delight we have snatched from the hands of our grandparents"] (*Fotografía de los colores* 18). The methods of science neither came out of nowhere nor would they disappear any time soon. Invention and innovation had been documented and esteemed by previous generations, existed in current generations, and held promise for future generations. Cajal reveled in scientific conquest (the triumph of the disciplined mind over ignorance), and marked a transition from the interpretive and artistic, the painterly image, and the hopeful conjecture to the illumination of microscopic and photographic image used to critique as well as create.

In an introduction to his manual on color photography published in 1912, Cajal saw chromatic photography—as did Benjamin when he wrote of chromatism beyond the stark tones of the black and white—as a mode of visual transmission that changed the experience of the viewer as much as it reflected the positioning of the viewer toward the image. At the age of sixty, Cajal wrote with almost youthful enthusiasm about the potential power of the photographic image. That changed relationship between observer and time hinged on the fixed photograph: "¡qué dicha sería poder contemplar, sin los afeites y convencionalismos de la pintura, siempre aduladora y esquemática, las juveniles facciones de nuestras madres . . . !" ["What a pleasure it would be to be able to contemplate, without the artifices and conventions of painting, always flattering and simplifying, the youthful features of our mothers!"] (*Fotografía de los colores* 18). There is a consciousness

of his own aging that tinged this longing observation on the passage
of time, yet it also drew attention to the experience of contemplation
itself as it does to the inviting traces found in photographs. The mode
of the painterly reflected the brushstrokes of the artist, while the camera
presumably produced (or reproduced) an image of a different sort, one
that accompanied the sentiment of rupture brought by modernity's
attempted break with the past. In the details of photography appear
flaws, nuances, facial tics and gestures, glances, and, above all, time
made material, visible, and concrete. Maybe what was recognized in a
painting now looked very distinct, some hidden nuance having been
brought out by the capture of an image on glass, metal, or paper, or
some flaw seen for the first time. That image may materially preserve
a moment into which the observer may interpellate him- or herself,
but it presumptively challenges and destroys traditional forms of perception—of face, space, and time—which, always the scientist, Cajal
proposed to capture with the instructions provided in his volume on
the best methods of color photography. The subtitle of the volume—
scientific bases and practical rules—acknowledges the attributes of the
scientist and those of the novice in equal terms. His photos simultaneously delve into the realm of the microscopic in search of understanding the mechanisms and architectures of the natural world, and into
domestic spaces challenged by modernity's changes. What better for
the production and reproduction of images than the most intense and
exact color, movement, contrast, and relief, just as he had worked on
all his life, in the realm of microphotography and stereoscopy?

Having internalized the melancholic lessons of 1898, and not
asserting the recapture of some past moment of grandeur or glory but
forging ahead "*de sorpresa en sorpresa*," Cajal instead recorded image after
image. He documented people, places, and his own work, aware that
new methods would come along and that each moment of progress
was one of pure contingency. He did not appear to posit a return to
something now lost but to an attitude of discovery that had been lost.
Although many Regenerationists, aware of Spain's failure to keep pace
with the social and cultural changes in the rest of Europe, "believed
that Spain's problems were essentially moral" (Ross 41), concerns over
land reform, public finance, corrupt politics, uneven industrialization,
and the fate of the workers' movements coexisted in diverse and even
contradictory measure. Cities grew as did skepticism over political solutions to the nation's divisions, and Cajal confessed that while he listened

to fellow scientists in various *tertulias* he turned in some exasperation to science rather than any political schema. When Cajal wrote of the quixotic spirit of adventure lost from Spain's most recent history, he evoked little of tilting at windmills. Instead, he sought more of a courageous spirit to confront the world as it is found then proposed how it might soon be constructed differently. For future generations in Spain who would inherit the debates to which he saw no end—to them, Cajal wrote in Spanish, whether about scientific topics or photographic ones—he summarized previous developments in his work, in color photography, and in new methods that promised to produce faster, clearer, and brighter results. That the products would also last longer did not hurt either. He frequently closed his memoir entries with references to future generations and to what he foresaw as exciting modern scientific work he would not live to see.

The term *modern* science sounds to our ears like a reference to some sort of empirical testing, putting theories under scrutiny and coming to results that prove hypotheses with a preponderance of evidence. Even popular accounts of the use of the scientific method have elicited doubts about the adequacy of some sort of procedure to produce a shift in the acquisition and integration of the natural world with each momentous innovation in technology. In the seventeenth century, the mathematician John Wallis, among the precursors of the Royal Society of London, could include in the "sciences" mathematics, philosophy, "Physick, Anatomy, Geometry, Astronomy, Navigation, Staticks, Magneticks, Chymicks, Mechanicks, and Natural Experiments" (cited in Brooke 55). But the "sciences," understood now as differentiated from theology, proposed a reconsideration of the past, of the ways of looking that belonged to the old order, to mark a cultural shift toward a reliance on reason aided by and through the senses. In particular, emphasis on vision and the lens to investigate and represent nature as both an active (the experimenter or willful self) and passive (restrained, observant, scientific self) object of cognition arose. Then photographic technology aided in hypothesizing an alternative standard and means of representation. The inquisitiveness, stubbornness, and challenge to authority that led Cajal to pursue scientific research, and the curiosity that drove him to avidly pursue photography, offer evidence to confirm the unique traits that were used to spur the promotion of this scion of science as a model of what would be required of Spain for entry into cultural modernity. Cajal's turn to drawings rather than language in the 1889 presentation of his findings using the powerful lens of the Zeiss microscope confirm on a microcosmic level the macrocosm of a community of modern scientists and intellectuals. His French and German being less than useful in communicating his discoveries, Cajal implemented his presentations to German biologists during the session with slides and drawn versions of cells. This visual medium crossed cultural borders and conquered linguistic obstacles. The community's shared language of science was that of the human eye.

Regarding experimentation and the scientific method, Brooke summarizes:

> there have been so many definitions offered by philosophers, and by scientists themselves, that it would require another book to consider them. Many refer to some singular, unique "scientific method" to which exemplary science is supposed to conform. But, as William Whewell, Cambridge philosopher and the first to coin the word "scientist" in the 1830s, observed almost a hundred fifty years ago, the *history* of science already showed that each new branch of scientific inquiry had required its own distinctive methodology. And that very process of increasing differentiation reflected a more fundamental change in the meaning of science—from when it had referred to all knowledge and when theology was "queen of the sciences," to its more modern connotations of empirical investigation and high specialization. (6–7)

A plurality of distinct methods, a specialization of types of inquiry, and a turn toward empirical information gathered from the senses by observation, experience, or experiment coincides with what Jürgen Habermas has described as the project of modernity: "Its project . . . is one with that of the Enlightenment: to develop the spheres of science, morality and art 'according to their inner logic' . . ." (Foster, xii). Beginning in the mid–nineteenth century, scientific inventions and innovations contributed to a possible shift in the acquisition and evaluation of these processes, as well as to the notion of an inner logic of disciplinary conventions. As Cantor and Brooke conclude: "there are fundamental ideas peculiar to each science" (139) and, therefore, conventions related to discrete fields that, while addressing specific concerns, might unite in a more encompassing vision of knowledge.

The grammar of histology and the grammar of photography are two codified sets of conventions that combine in the activities of Cajal as both observer and investigator; they also overlapped in scientific investigation, the focal point of his life. Despite his proclamation of breaking the codes of painting with the use of photography, there is a grammar of this technological innovation linked to the times, a set of structural rules that govern the composition of images. Cajal inherited norms and practices of composing photographs, of setting up a still life, of preparing a slide, of writing up a report. But he was never content with that, and he constantly pursued faster, truer, more accurate results, or the portability of equipment that took photography into the world. Challenges that might arise from what had not yet been made available to the human eye were not to be avoided but accepted, using sophisticated lenses of all types as light-providing intermediaries between the eye and the brain.

The increasingly potent and convincing norm of objectivity for the observation of all phenomena at all levels of perception began to take root in Spain, as in the rest of Western Europe, around 1860, following technical developments in the scientific equipment used to conduct experiments related to the description (how) and explanation (why) of what could be known about the natural world. Into this context of investigation and inquiry, the rapid translation and publication of Daguerre's manuals on the photographic process into Castilian may be added.[3] With these works, one might capture the workings and inhabitants of the world with a more radically modern—seemingly unassisted by any intervention—vision supported by the lenses of modernizing technology. These included photography, stereoscopy, photomicrography, and compound lenses whose enhanced power of sight demanded an equally enduring indelibility of the images produced. The more could be distinguished, the more one could study and comprehend. The material devices and trappings of modernization that accompanied a transition from the traditional in the realms of economics and industrialization might not necessarily have brought an epistemological change—an intellectual culture of modernity, or the notion that change had brought about a rupture with a past worldview and its quest for knowledge—along with them. It is one thing to find the construction of railways, telegraph lines, urban centers, the establishment of university chairs in science, and photographic studios; it is another to consider the technologies of modernization as valuable sources of knowledge. Walter Benjamin did not reckon with the objectivity but with the "ecstatic" aperture of the structures of experience afforded by the camera, "the potential for infinite transformation opened by a technologically informed experience [which] can either be affirmed, leading to constant innovation in the subject or reality, or refused in the regressive use of technology to restore distance . . . and permanence—i.e., monumentality" (Caygill 105). While in later decades of the twentieth century, the Spanish state used the power of cinematic technology to proclaim and solidify its claim to rule, Cajal found in the laboratory and the darkroom the potential to put theories to the test as well as to discover the surprises of the unexplored.

Of such values, Daston and Galison clarify that objective observation was not valued always but that

> Scientific objectivity has a history. Objectivity has not always defined science. Nor is objectivity the same as truth or certainty, and it is younger than both. Objectivity preserves the artifact or variation that would have been erased in the name of truth; it scruples to filter out the noise that undermines certainty. To be objective is to aspire to knowledge that bears no trace of the knower . . . only in the mid-nineteenth century did scientists begin to yearn for this blind sight, the 'objective view' that embraces accidents and asymmetries." (17)

Nature seen as divine creation and mystery, or nature seen as a collection of harmonious typologies met up with a modern quest for firsthand knowledge about the world. Cajal recorded that his early experiments on cadavers provided him direct experience of *"cosas objetivas y concretas, acogía con ansia el pedazo de maciza realidad"* ["objective, concrete things, . . . I anxiously accepted (all) fragments of solid reality" (*Recuerdos* I, ch. XVII). These he could examine with "a passionate commitment to suppress the will" (Daston and Galison 143) and be "truthful" in his conclusions. The virtue of objectivity stood squarely at the center of Cajal's groundbreaking work, and as a clear presence in his qualms about the objectivity of the scientific practices and conclusions as practiced by co–Nobel-winner Camillo Golgi. But it also formed a substantive part of his simultaneous fascination with photography as chemical process and as a visible product. The two fields were not just contiguous in the narrative retelling of his life; instead, they overlap and interrupt one another, indicating how thoroughly and deeply the shared values penetrated both.

As he looked back at his early experiments with photographic plates and with "chromatism" or color photography, Cajal recalled his great "enthusiasm for the art of Daguerre" and, adding in the same breath, for obtaining an appointment as director of the Anatomical Museum in Zaragoza. At eighty, Cajal had the perspective to comment on the advances as well as the manipulation of such processes. He wrote: "Practico el arte de Daguerre desde los dieciocho años y conozco todas las tretas, trampantojos y abusos que con ella pueden cometerse. Y me son familiares las artimañas del cine" ["Since the age of eighteen I have been practicing the art of Daguerre and I know all of the artifices, tricks on the eye, and abuses one can commit with it (photography). I am also familiar with the clever deceptions of the cinema"] (*El mundo visto a los ochenta años*, 157). Both photography and the position in Zaragoza signaled an end to his period of recuperation from the malaria and lung disease he contracted during his service in Cuba in the medical corps. Cajal retold in detail his diet in the monastery of San Juan de la Peña (an architectural structure he photographed often during his confinement), the tranquillity and "affability" (267) of the consumptives around him, the distraction of taking photographs and preparing the plates, "[the obligation to] take continual exercise" in search of subject matter, and "the daily solution of artistic problems" (268) related to the quality of the images. Each was a challenge for this curious artist and scientist in need of a focus, a way to pass the time and to refine his technique. Cajal turned to technology at the service of art in his insistent camerawork as he recovered. At the end of this lengthy passage about the sanatorium, he concluded with an apology: "Begging the reader's pardon for the foregoing digression on photography" (269). This is followed by a return of the narrative to histological studies and the laboratory. Such digressions appear frequently in his memoirs, and rather than detract from scientific discourse, they complement them as "fragments of solid reality" related to the chemical developments in photography he explored in the laboratory and in the field.

The overtly contentious relationship between 1906 Nobel Prize cowinners Golgi and Cajal was predicated on claims of exactitude in the reproduction of images and the conclusions drawn from direct observation. It combined the Golgi stain or method of making microscopic structures visible with Cajal's development of the neuron doctrine and an improved Cajal stain. The work of the two notable scientists marked a transition from nerve cells traditionally observed under the microscope then reproduced in light of conclusions that smoothed out anomalies to the "undistorted sight" (Daston and Galison, 116) proposed through the use of the lens. Cajal saw his own depiction of the nerve cells, ending in spaces and not conjoined cells, as direct and faithful confirmation of his thesis. The images told different stories, but the sense of a need for "right depiction" contributed to the tensions in representation between the earlier paradigms of observation and the new code of objectivity that reproduced difference, anomaly, and exception. Irregularities in structures, and the potential for variability, contributed to an "epistemic instability" (Daston and Galison, 50) seen by some as a threat to the "reasoned image" (42) that maintained underlying natural harmony and permanence. For Cajal, harmony was not the issue. In his quest for "rightness" of depiction, Cajal proclaimed his absence from personal subjective intrusion. Yet his value as a scientist who would put Spain on the modern European map was always present, requiring some sort of acknowledgment of his presence in every experiment. Of course, artistic skill and intuition guided the production of an image, but the self could only be celebrated in the figure of the scientist, not the procedure. If "both artistic and scientific personas spawned heroic myths, albeit complementary ones" (Daston and Galison, 146), it is seemingly rare to find the two in one. Cajal found both rightness and beauty in the authenticity (right seeing) of an aesthetic photograph and the authenticity (again, rightly seen) of a scientific image that responded to different conventions: expression, in the first case, and discipline, in the second. The active scientific experimenter—in the chemistry of the photographic process as much as in the microscopic vision—and the passive observer could function in both spheres and indeed work in complementarity. Martino Rossi Monti calls attention to the fact that "scientific images are often considered, by those who create them as by those who contemplate and use them, to be works of art" (280). An album of scientific drawings or slides can coexist alongside an album of photographs, both linked by the use of technologies related to the camera, the microscope, and the lens.

The rise of a cult of the lens that encompassed the invention and development of the Zeiss superb-quality "wide open" microscope

(1840s–80s) whose aperture allowed for extremely bright and detailed images, stereoscopic photography, the autochrome color process, calotypes, and the Daugerrotype offers instructive material proof of the value adjudicated to the "calibration of the eye" (Daston and Galison 44). This occurred in both scientific and popular cultural circles in Spain between 1839 and the early twentieth-century. Diana Saldaña surveys the rise of photographic interest in Spain among the cultural and scientific elite and nonprofessional circles alike, finding Spain a fertile market for foreign Daguerreotypists in the 1840s,but an uneven market at best. But by 1868, Spain was a nation in the thrall of photography as "a technology that symbolised modernity" (Saldaña 1326) in both process and product. In this respect, certain segments of Spanish society were not that different from the Victorian era in England (1837–1901), in that the parlors of wealthy families were spaces for the enjoyment of both microscopes and cameras as technological sources of entertainment. Stereoscopes sat on parlor tables. In Spain, the documentary aspect of these technological inventions overcame their pleasure value. Cajal was never a dilettante, so his adoption of photography as an intellectual activity went far beyond the parlor and the lightheartedness of family entertainment into the realm of observational skill and the recording of a permanent image. His camera was the instrument that recorded the conjunctions of family life and scientific pursuits when he composed still life settings using fruit, pottery, patterned cloth, and laboratory equipment. One can see in the eyes of his wife and children that the photographs he had them pose for required them to stop their activities to sit for the portraits. For him, these were not moments of entertainment but the scientific testing of development techniques and color definition. Cajal's practice of photography was motivated as much by science as by an interest in collecting images of the family.

Strengthened relationships among official court photographers from Isabel II (1858) on, the Spanish press, official topographical photographers to document the planning and construction of modern buildings, visual diaries of travelers and literati from abroad, and the expansion of a photographic industry within a nascent capitalist modernity in Spain led primarily in two directions.

The first, the domain of aesthetic or art photography, can be understood as a continuation of a romantic vision that linked the promising wonders of technical and scientific modernization with a previous aesthetic of an inimitable and uncommon Iberian sense of beauty. The construction of new monuments for modern times replaced the inherited exoticism of Granada and its Moorish past without changing the tone (and tint) of fascination, this time with Spain's recent and somewhat tardy entrance into in the realm of industrialization. Debates around photographic pictorialism—a studied manipulation of the image to achieve specific aesthetic effects and stir affect in the eye of the observer—contributed to the recording of Spain's architectural treasures through a subjective enhancement of their aesthetic qualities for the foreign visitor. Kurtz writes of Spain as a land that time forgot that attracted curious travelers "lured by the fabulous—and, it was feared, disappearing—remnants of a glorious past" (1). Nineteenth-century photographers from the United States, England, France, and other European countries made excursions to document such exotic locales as antidotes to European modernity. They adhered to the codes of subjective experience over objectivity. Albums, individual and family portraits, *cartes-de-visite* or calling cards, group photos, travel snapshots, army encampments, scenes of war, and political propaganda all contributed to "*un canto del progreso*" ["a paean to progress"] (Lara López 5) that popularized and democratized the art of photography which, certainly, employed science to record and develop the image. Cajal provided photographs from an insider's perspective, rather than from the point of view of outsiders among the ruins of a nation cautiously entering a tenuous modernity. He revisited and recorded the crumbling walls of his childhood home, as well as the parks, streams, and outskirts of the city. The bustling streets of Madrid contained fountains, new constructions, and public spaces, all accessible through portable cameras and faster development of images. Cajal recorded it all, from a simple home laboratory to a professional academic setting. His collection of images produced during his travels covers a diversity of venues, some related to portraits and self-portraits, others to relatives posed amid scenic natural overlooks, as well as to old and new architectural landscapes of Spanish cities and towns. From Cuenca to Zaragoza, from Barcelona to Madrid, Cajal focused his camera on the cultural artifacts of both construction and destruction, on centuries-old monasteries such as La Cueva in Aragón, and the new builds of La Gran Vía. Never a professional photographer by trade, Cajal nevertheless embodied the professionalization of photography as a national endeavor, as well as the professionalization of photographic methods as chemical and laboratory-related endeavors with increasingly advanced and advantageous development processes and optical inventions.[2] More about his work in the era that issued in photography as a primal aspect of modern technologies will be elaborated on in chapter 3.

The second direction is the two-way street between scientists as active and supportive disseminators of photography in Spain and, in turn, that same technology that "was to become an indispensable tool for science as technology continued to develop" (Saldaña 1326). Following in the footsteps of the tradition of Alexander von Humboldt earlier in the century, Spanish expeditions to South America in the 1860s required precise documentation of their discoveries, so there was nothing more fitting and scientifically satisfying than the claims of "mechanical photography" (Daston and Galison 133) to record their findings in collections of specimens and images. The same held true for José Monterrey and Warren de la Rue's photographic registering of the solar eclipse of 1860 that so enraptured Cajal when it appeared in local news reports, or the use of photography for medical purposes in Barcelona during the 1870s.

The rapid growth of itinerant portrait photographers and, later, photographic studios marked a golden age of photography in Spain up through the national crisis of 1898. By that time photography had become thoroughly rooted in the popular imagination. The quality of photographs, as well as improved pricing and mobility, gave amateurs an advantage, especially with the advent of bromide gelatin dry plates and the introduction of the Kodak box camera to Spain in 1888. Not his only camera, by any means, Cajal's Kodak—noted as a staple of his travels through the Levantine region (*Recuerdos* II, ch. III)—accompanied him as he crossed mountain ranges, recorded landscapes, and breathed the fresh air among the palms and pines. Clubs and associations such as the Photographic Society of Madrid (later, the Royal Photographic Society), founded in 1899 and over which Cajal presided for many years, established once and for all the importance of visual culture as both art and science, from family portraits and social documents, to photomicrography and microphotography. By the time 1860 had dawned, photographic techniques had become easier to master and emulsions less dangerous. In this atmosphere, Cajal contributed a refinement of the chemical development processes both in the speed in which an image appears, and in the quality of light-and-dark contrast. Beaumont Newhall, photographic curator at the Museum of Modern Art in New York and curator of a 1937 exhibit commemorating the first century of photography, recapitulates the aesthetics of the photograph as well as the chemical processes that responded to an increasingly competitive marketplace: "it located two main traditions of aesthetic satisfaction in photography: from the optical side, the *detail*, and from the chemical side, *tonal fidelity*" (Phillips 19). These two characteristics, or qualities, of the photograph as process and as product, broaden the notion of what constitutes an artistic image. Detail and fidelity are relevant to portraits, landscapes, and photomicrography or microphotography.

The deployment of the practices and instruments of scientific inquiry raised questions about the complexity as well as the beauty of what was being looked at in addition to the aesthetics of scientific pursuits in and of themselves. Doing science held a beauty all its own, and the results of cautious observation held a special fascination. Different media represented new ways of seeing and being seen through "devices that could exist anywhere on a continuum that ran between, but always blended, spectacular entertainment and educational intent" (Hand 927). The artistic photograph existed alongside the scientifically perfected one, distinguished by the educated eye behind the lens through which the representation of nature was recorded. The first was ruled by affect and desire, while "the divided scientific self, actively willing its own passivity" (Daston and Galison 146) controlled the second with a declared desire for the elimination of all personal intervention. While hard to find such objectivity in the content of the photograph, Cajal nevertheless could turn to the accuracy of processes to fix faces, landscapes, and structures on a surface. Working from his home in a room designated as his laboratory, employing the devices of both microscope and camera, Cajal himself may be the best bridge of this focus on the visual as simultaneously scientific and imaginative, as a potentially objective device and aesthetic object. The scientist himself referred to his earlier drawings in the dissection room in terms that we could use for his immersion in photography. Instead of skills deployed to look at death and destruction, anatomists could envision their art at the service of life, and photographers might do similarly as they fixed solutions that reflected "*el admirable artificio de la vida*" ["the marvelous artifice of life"] (Cajal, *Recuerdos* I, ch. XIX). Driven by curiosity and imagination, by the puzzle of the human body as healthy entity or site of pathologies, and as an impulse for human interactions, Cajal's anatomical catalogs and drawings are the staples of many generations of microscopists and neurologists. His work as a photographer has been glossed over but much less explored.

Cajal exemplified both the spirit of the collector and classificatory taxonomist, and that of the image-maker as a recorder of what he has observed and made into a compendium of knowledge about the visible world, including the microscopic one. Figure 1.1 is a photograph of Cajal's cabinet of prepared laboratory slides from all of his laboratory experiments, arranged in drawers and labeled with such care so as to preserve an organized and accessible record of his work in the field of histology. Available at arm's reach in organic arrangement and classification, the labels and drawers catalog information that could be built on, reconfirmed, or reconsidered. But before the cataloging of these slides, Cajal's earliest collections were material objects drawn from the natural world around his childhood home that he gathered on field walks.

Looking back some fifty years later, from the vantage point of the decades between 1901 and 1917, Cajal recalled that "frisaba ya en los trece años, cuando di en coleccionar huevos de toda casta de pájaros, cuidadosamente clasificados . . . En estos caprichos no entraba para nada el interés gastronómico ni la vanidad del cazador, sino el instinto del naturalista." ["When I was about thirteen I took a notion to collect eggs of all types of birds, carefully classified . . . In these amusements there entered nothing of gastronomic interest nor the vanity of the hunter, but only the instinct of the naturalist"] (*Recuerdos* I, ch. III). This collection of 1865, kept by the young naturalist-in-training in a carefully labeled and compartmented cardboard box before the unexpected destruction of some of the eggs in the heat of August, was enhanced by non-interruptive methods of collection and disposal. His connection to the natural world was as unintrusive as it could have been; he says it was motivated by "un sentimiento de clemencia" ["a feeling of mercy"] (*Recuerdos* I, ch. III). The live capture of nestlings with "*besque* o liga, *lienas* con hoyos hondos, la red, etc." ["a thick paste of bird lime, *lienas* with wide holes, nets, etc."] to observe their growth is followed by their gentle release back into the wild "a sus nidos y a las caricias maternales" ["to their nests and to their mothers' care"] (*Recuerdos* I, ch. III). There is nothing in Cajal's studies of nature of the specimen preparation and taxidermy, for example, of John James

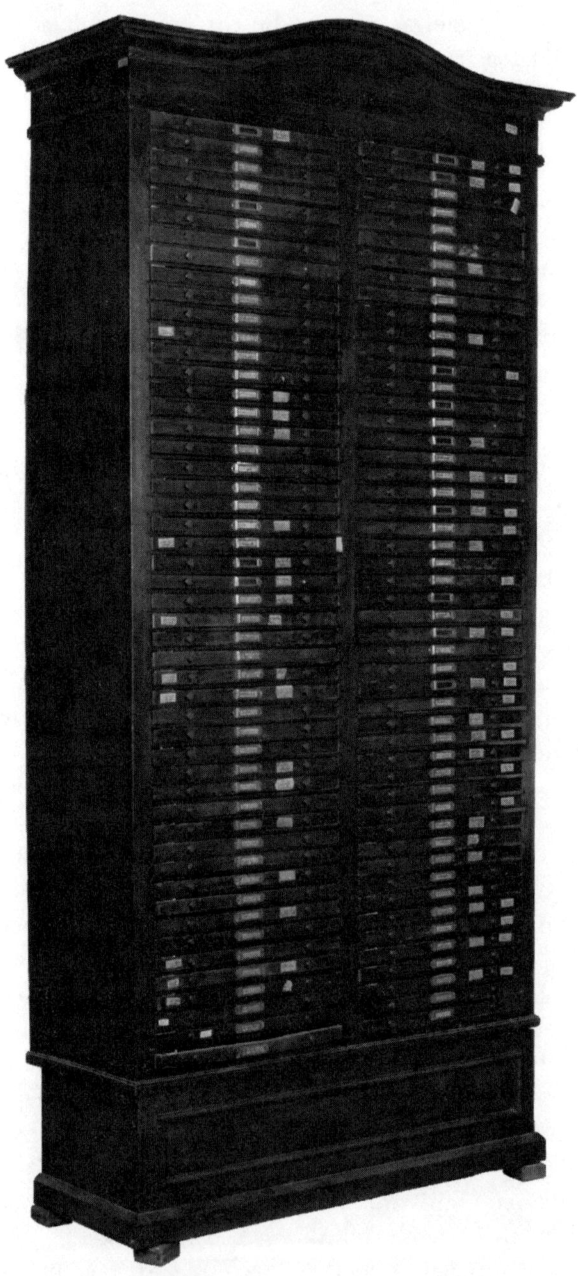

Figure 1.1. *Armario de preparaciones* [*Cabinet of slides*]. Santiago Ramón y Cajal. Legado Cajal (CSIC). Instituto Cajal. Madrid, Spain. Used by permission.

Audubon's bird collection. Audubon (1785–1851) developed his own methods for drawing birds, a process far from the ends of Cajal's collecting for classification of functional structures, and how to distinguish between correct and errant forms. Audubon killed birds using small shot, then reconstructed them in flight or repose using fine wires to prop them into a natural position. His requirement of natural poses differs from more common methods of ornithologists of the time who prepared the specimens into rigid, lifeless poses. In his publication *Birds of America* (1827–38), Audubon's paintings of birds are set true-to-life in their natural habitat, a habitat to which they will, however, never return. They might be represented interacting with other specimens, surrounded by an appropriate environment, but the poses are frozen pauses in the lives of the creatures. Accurate as they can be, the detailed drawings have stopped life in midflight. While Audubon based his paintings on extensive field observations, the young Cajal brought the natural world, still alive, into his home. Unobtrusive observation ruled his treatment of the natural world. His later images of that world will be as insistent and curious in their technique and composition. Whether birds or humans, all appear in the midst of being observed by the scientist and the public alike within a community.

Cajal found it sufficiently suggestive to observe the young fledglings. He also found scientific knowledge latent in the sunlight broken into the optical phenomenon of a spectrum of colors that streamed through the schoolhouse window, a result of the tower of his schoolhouse being struck by lightning, and the implications of the solar eclipse of 1860. Unlike the magical interpretation of an eclipse during colonial times, or superstitious connotations linked to such a cosmological event, Cajal was fascinated by the observable phenomenon and how light was affected by it. In addition, from the vantage point of old age, Cajal recalled the diverse effects of the solar eclipse on the natural world as it interfered with the regular intervals of night and day—darkness and light—as well as the event serving as a turning point in his own observation of the world. Cajal wrote: "Se comprenderá fácilmente que el eclipse del año 60 fuera, para mi tierna inteligencia, luminosa revelación. Caí en la cuenta, al fin, de que el hombre desvalido y desarmado enfrente del incontrastable poder de las fuerzas cósmicas, tiene en la ciencia redentor heroico y poderoso y universal instrumento de prevision y de dominio" ["One can easily understand that the '60 eclipse was, for my youthful and tender intelligence, a luminous revelation. I finally realized that man when seemingly helpless and destitute faced with the unequaled power of cosmic forces, has in science a powerful and heroic redeeming instrument of prediction and control"] (*Mi infancia y juventud*, 47)]. The language of this brief excerpt exposes the awe of such a sight and the potential of scientific observation and subsequent knowledge to explain, and even harness, the power of the natural world. Cajal used metaphors—"la casta Diana acudió a la cita" ["chaste Diana arrived right on time for her date"] (*Mi infancia y juventud*, 46) for the eclipse of the moon—and adjectival phrases to represent the impact of the obscuring of the heavenly body, the rainbow of light through a prism, or the portents of the modern railroad, the cinema, and photography. His youthful enthusiasm, boyish and innocent, is rendered relative by the old man who lamented, but now understood, his own shortcomings. What were early, "*vulgares experimentos*" ["common experiments"], "*la ciencia de las maravillas*" ["the science of marvels and wonders"], "*impensados descubrimientos*" ["unexamined discoveries"], or "*fantasmas luminosos*" ["luminescent ghosts"] (*Mi infancia y juventud*, 67) were contextualized as potentially promising but lacking precision, instinctive and simplistic. The poetry of natural phenomena never disappeared from Cajal's discourse. Knowledge about laws of nature, and scientific observation and annotation, came later with experience.

Observation rather than conjecture stimulated the inclusion of many species in his ornithology box, as it does in his notebooks. The young Cajal's recorded lists of local birds, observed and preserved in all their variation and variety, seems to have trained the future observer of natural objects, whether under the microscope or from behind the camera lens. His "treasure" was a collection of specimens gathered into an aviary compilation of similarities and differences, norms and variations, each labeled and designated as specifically as possible. The repeated activity of collection and classification—along with curiosity, resourcefulness, obstinacy, a passion for objectivity, and imagination—were stimulated by Cajal's early training as well as by the scientific example of his father. In that paternal figure he recognized the source of his own will (to organize, to understand, to know), the stubbornness he shared, and "*la convicción de que el esfuerzo perseverante y ahincado es capaz de modelar y organizar desde el músculo hasta el cerebro, supliendo deficiencias de la Naturaleza y domeñando hasta la fatalidad del carácter*" ["the conviction that deliberate and insistent effort is capable of molding and organizing from the muscle to the brain, compensating for the deficiencies of Nature and domesticating character flaws"] (*Recuerdos* I, ch. I). The visible world around him would subsequently yield—or open up to—a vast array of both smaller and larger phenomena on which Cajal could focus. These he arranged in carefully labeled cabinets and drawers, categorized by genus and species according to taxonomic guidelines. Much later, he kept his slides in similar order.

The anatomical sketches and watercolors done at his father's request constituted a second example of Cajal's mastery of scientific classificatory practices and dedication to them. An outgrowth of his artistic interests in his youth, such as the project to

> reproducir en grueso álbum todos los matices variadísimos ofrecidos por los objetos naturales, ejecutando una especie de diccionario pictórico, donde, a falta de nombre, cada color complejo figurase con número de orden . . . añadíale la imagen del objeto correspondiente. Era algo así como la conocida gama cromática de Chevreuil (que yo ignoraba entonces), . . . aparte los tonos simples más o menos saturados, el producto de la mezcla de todos los colores, incluyendo naturalmente el blanco y negro.

> ["reproduce in a thick sketchbook all the various shades represented by natural objects, making a sort of pictorial dictionary in which, though lacking a name, each complex color would appear with a number and the corresponding object . . . It was something like Chevreuil's chromatic color scheme (which I was not aware of then) . . . besides the simple colors in various degrees of saturation, the product of mixing all the colors, naturally including white and black."] (*Recuerdos* I, ch. XII)

Cajal's interest in hue and tone surfaced again in his work on color photography. Subsequently, Cajal turned to the study of the human body. The original meaning of autopsy—seeing with one's own eyes— came into play as Cajal plumbed "the marvelous workmanship of life" (169) and debated the value of becoming a physician or a surgeon. He found in each field of medicine the sort of "inner logic" Jürgen Habermas associates with the modern, with internal pathology (the domain of the surgeon) related to a "*ciencia contemplativa*" ["contemplative science"] and external pathology being "*como ciencia de acción y de dominio*" ["like a science of action and control"] (*Recuerdos* I, ch. XVII), which could intervene in all processes. Both require observational skills. Cajal uses another analogy to present the rivalry between the disciplines: "existirá siempre entre el cirujano y el médico la misma relación que entre el diplomático y el caudillo. Quien persuadiendo triunfa, granjea opinión, no libre de envidia; quien triunfa combatiendo, tiraniza hasta la envidia misma. Tras éste corre desalada la gloria; aquél suele perseguirla sin alcanzarla" ["Between the surgeon and the physician there will always exist the same relation as between the diplomat and the military commander. He who wins by persuasion earns esteem to be envied, while he who triumphs in battle tyrannizes envy itself. Glory follows the latter quickly, the former may pursue it without ever earning it"] (*Recuerdos* I, ch. XVII). He discovered in surgery a supremacy that divided the science of medicine, as he saw theory and practice in competition for social recognition and evaluation.

Ascribing a preference solely for the objective sciences over the theoretical and the abstract to his adolescent years, along with hyperbole and other "*aberraciones del gusto*" ["aberrations of taste"] (*Recuerdos* I, ch. XIII), he nevertheless found in the fundamentals of anatomy a chance to view and then reproduce detailed observations of the recesses of the human body. Theory in that case led to practice and observation, not the distancing of one from the other. Through repeated efforts, Cajal satisfied his thirst "*de cosas objetivas y concretas*" ["for the objective and the concrete"] (*Recuerdos* I, ch. XVII), thereby avoiding the "*crimen didáctico*" ["didactic crime"] (*Recuerdos* I, ch. XVII) of secondhand study. Tracing from paper could not provide the understanding of nature that direct study would, and his access to human cadavers reveals yet another facet of the value of observation. In the laboratory, Cajal and his physician father pored over human bodies:

> Nada esencial quedó por reparar en la morfología interior y exteriorde cada pieza del esqueleto. Bien miradas las cosas, mi fervor anatómico constituía una de tantas manifestaciones de mis tendencias; para mi idiosincrasia artística, la osteología constituía un tema pictórico más . . . Sentía, además, especial delectación en ir desmontando y rehaciendo, pieza por pieza,

el reloj orgánico, y esperaba entender algún día algo de su intrincado mecanismo. (*Recuerdos* I, ch. XVII)

[Nothing important remained unobserved in the internal or external morphology of each piece of the skeleton. If things are looked at in their true light, my enthusiasm for anatomy was just one of the many pieces of evidence of my interests; as for my artistic idiosyncracy, the study of bones constituted one more subject for pictures . . . Moreover, I felt a special delight, in taking apart and putting together again, piece by piece, the organic clock, and hoped someday to understand something of its intricate mechanism.]

Not quite as intricate perhaps as microscopic cells, but arguably just as fascinating material for observation, social reality, the collection of diverse human bodies and countenances forming a family or a society, was studied by Cajal through the lens of the camera. He recorded in his collection of photos, again piece by piece, the parts of the whole to understand their inner workings and their social interaction. What his writings show about a confluence of the two arenas of expertise—anatomy and artistry—was that he pinpointed the nervous system and the brain as the indisputable centers of all facets of human life.

The appeal of these biological structures connected them subsequently to interrelated studies of behavior, pathology, and psychology. Laín Entralgo and Albarracín recount that during his first years as a teacher in Valencia, Cajal organized a Committee of Psychological Research in his home. This he dedicated to the "experimentación y terapéutica de neuróticos. . . . A mi consulta acudían enjambres de desequilibrados y hasta locos de atar" ["experimentation with and therapeutic treatment of neurotics. . . . Swarms of people showed up at my office, from the unbalanced to the stark raving mad"] (*Mi infancia y juventud* 87). Inside or outside the human body, within other tissues or within the brain, every aspect of life was open for observation. Cajal created note cards and notebooks with data on everyone who turned up on his doorstep just as he archived slides. Cajal was the cataloger of everything visible and, as Llinás concludes, he "personifies, above all, the belief that we actually can understand the nervous system, which represents, more than anything else, the very nature of what we are"

(80). To queries about what Spain and its inhabitants might "be" after the turn of the century, Cajal acquired an all-encompassing collection of evidence to advance toward some possible conclusions.

The photographic album illustrated a move outward from the traces of the life of the infinitely small seen in Cajal's "*álbum anatómico*" ["album of anatomical sketches"] (*Mi infancia y juventud* 27) and renderings of neurons. His journals published in the 1880s are also illustrated with lithographs. The photographic process encapsulated the dual focus of objectivity (empirical observation) and subjectivity (a privileged expression of the self held at bay by reigning scientific principles) through a dual "preservation of the artifact" by a "practiced eye" (Daston and Galison 17, 18). The lenses, slides, sketched images, perfected use of chemicals and dye processes, and photographs all produced both pleasure for the photographer—first as serious investigator of the natural world, and second as compiler of intriguing traces of the social world—and a record of cultural shifts in the composition and reading of the image. While Cajal categorized his use of the camera for purposes other than scientific investigation as a distraction from the rigors of scientific research, the distance was not as great as it might first appear. In his autobiography, he stated that the art of Daguerre was fundamentally a practical use of the physical principles of chemistry, a "marvellous application of science" (579) offering the opportunity to put theoretical hypotheses into practice. The two were considered fundamentally united: "To deprive oneself of the theory . . . is to disdain half of the pleasure of colour photography, which consists in testing experimentally the precision of the scientific principles. . . . The interpretation of the results obtained and the remedy for accidents and failures are to be found only in a clear understanding of the physico-chemical mechanism of each photographic operation" (579, 581). His overwhelming curiosity displayed since early childhood surfaced equally in the laboratory, the family sitting room (partitioned for impromptu use as a photographic studio), in burgeoning cities at the turn of the century, and among the woods, fields, and valleys of the Spanish countryside. As his scientific work promoted his presence at international events, Cajal became a photographer of other cultures, his students, colleagues, and young protégés.

In one example (figure 1.2), *Retrato de Silveria* [*Portrait of Silveria*] is a close-up black-and-white photographic portrait of Cajal's wife, Silveria Fañanás García, whom he married in 1879. She appeared quite

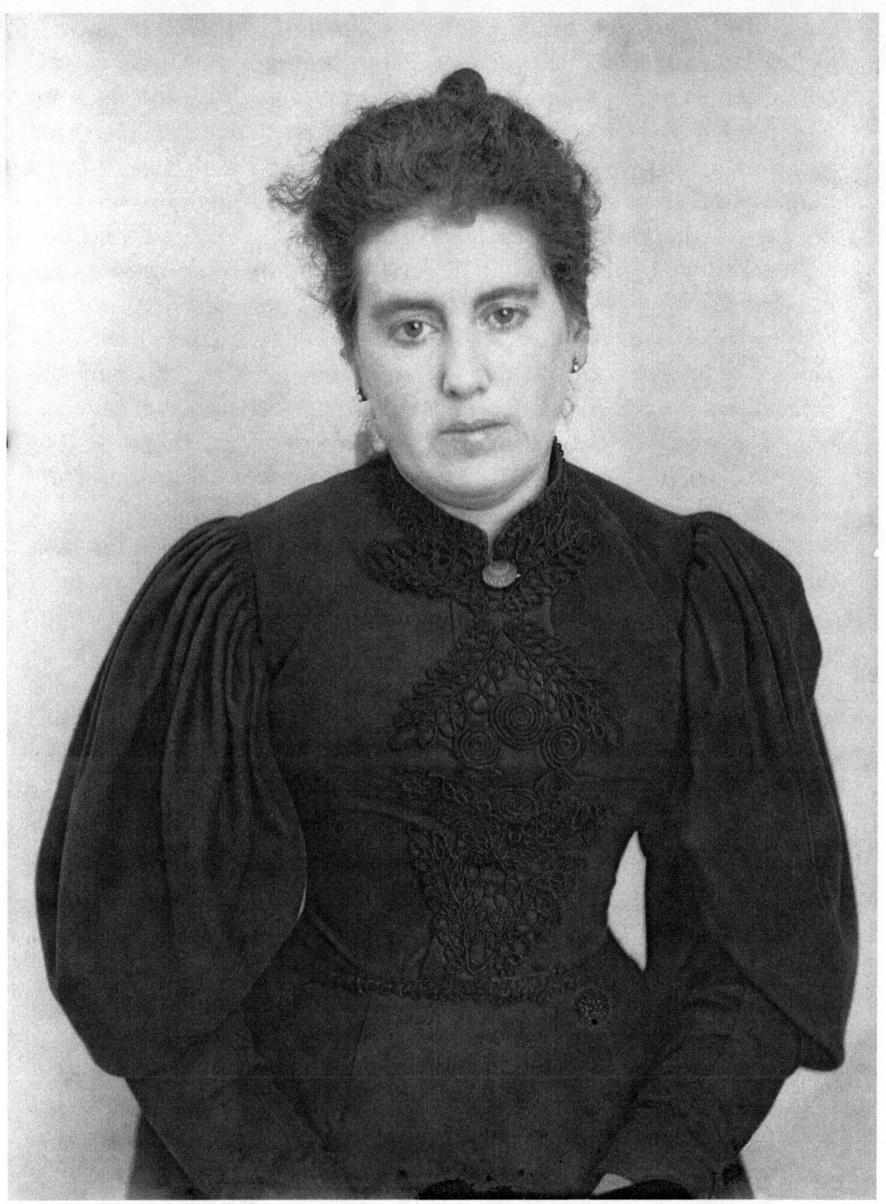

Figure 1.2. *Retrato de Silveria* [*Portrait of Silveria*]. Santiago Ramón y Cajal. Legado Cajal (CSIC). Instituto Cajal. Madrid, Spain. Used by permission.

frequently in his photographs, whether alone, by his side, or accompanied by several of their children. She is almost always looking off-camera or has her eyes closed. One reason may have been the length of time needed to expose the plate. Another, more enticing reason, could be a hint at her private thoughts even as she posed for his experimental camerawork. Here she faces the camera even if her eyes do not meet the gaze of the photographer. Posed in a fairly elegant dress with lace embroidery along the neckline, Silveria appears more composed than in other shots where she looks like she has interrupted her domestic activity to pose for the camera. She rarely makes eye contact but does accede to her husband's wishes for capturing portraits of family life with ever-new and innovative equipment and techniques of preparation. The observer cannot fathom what her thoughts might be, but Cajal imbues her image with the potential for imagining something other than domesticity and work.

Aside from being the subject of many such studies over the long duration of their marriage, and of standing in as a measurable image of the passage of time, Silveria also provided assistance with the laborious chemical aspect of his experiments in photographic development. During the evening hours, after he had completed his histological work in the laboratory, Cajal turned to the camera and to what he wrote of as the "*augusto misterio del cuarto obscuro*" [august mystery of the dark room"] (*Recuerdos* I, ch. XVII) for activities to fill the rest of his time. Wet collodion development of images on glass made portraiture possible, he believed, a fact that caught his attention by bringing into visibility an image where a few instants before there had been nothing to see. On this topic, Cajal recounts

> Todas estas operaciones produjéronme indecible asombro. Pero una de ellas, la *revelación* de la imagen latente, mediante el ácido pirogálico, causóme verdadera estupefacción. La cosa me parecía sencillamente absurda. No me explicaba cómo pudo sospecharse que en la amarilla película de bromuro argéntico, recién impresionada en la cámara obscura, residiera el germen de maravilloso dibujo, capaz de aparecer bajo la acción de un reductor. ¡Y luego la exactitud prodigiosa, la riqueza de detalles del clisé y ese como alarde analítico con que el sol se complace en reproducir las cosas más difíciles y complicadas, desde la maraña inextricable del bosque, hasta

las más sencillas formas geométricas, sin olvidar hoja, brizna, guijarro o cabello! (*Recuerdos* I, ch. XVII)

[All these operations astonished me tremendously. But one of them, developing the latent image by means of pyrogallic acid, positively stupefied me. It simply seemed absurd. I just couldn't explain how one could imagine that in the yellow silver bromide film recently exposed in the camera there might be concealed the germ of a wonderful image, able to be made visible under the action of a reducing agent. And then the marvelous exactitude, the richness of detail and that sort of analytical display with which the sun delights in reproducing the most difficult and complicated things, from the inextricable tangle of the forest to the simplest geometrical forms, without overlooking a leaf, a filament, a pebble, or a single hair!]

The combination of wide-eyed fascination and down-to-earth scientific curiosity led him to ask questions of the photographers who he found to be devoid of intellectual curiosity and driven to make money from the great "accidental" invention of Daguerre. Their "indifference to the theory of the latent image"—a comment Cajal finished with another exclamation point as if in absolute disbelief—became the start of a meditation on photography as part of a world yet to be discovered. He called these "enigmas, hidden properties, and unknown forces" that science would never exhaust but that lurked outside the realm of current understanding. Cajal seemed to understand the time of the modern as the beginning of a shift that would expand the horizon of scientists and regular citizens in unexpected directions, hopefully bettering human life in the process. Obstinacy and persistence were the qualities he reiterated as necessary to deal with such overwhelming possibilities, as well as to undo the myths of fate or chance still alive in the nineteenth century: "*nos rodea aún nube tenebrosa sólo a trechos rasgada por la humana curiosidad*" ["we are still surrounded by a dark cloud only rent here and there by human curiosity"] (*Recuerdos*, I, ch. XVII). Cajal proposed that the world still held mysteries, that "la ciencia, lejos de estar apurada brinda a todos con filones inagotables" ["science, far from being exhausted, affords everyone inexhaustible deposits of ore"] (*Recuerdos* I, ch. XVII). In very poetic terms, he positioned scientific invention and technological innovation at the crossroads of change whether achieved by perseverance or by so-called chance.

So to the language of scripture, and then to the language of the natural world, the mid-nineteenth century added a language and grammar of photography. The incipient photographic technology engaged and united the fields of art and science with equal vigor and passion, and provided particular structures of meaning. The process that captured the details of the invisible world—the world too small to be seen with the naked eye as well as the world the eye is not trained to notice—employed a mechanics that functioned to record, to inventory, and to memorialize. The collector of material objects could use this technology to record, the scientist to inventory and organize, and the man to distinguish the characteristics that identify himself and his surroundings in all their heterogeneity. It is not sufficient to examine Cajal's scientific activity without turning to his use of lenses, chemicals, and other elements of photography to introduce an album of images for consumption alongside his laboratory notebooks and sketches. The album as faithful recorder of the natural world (albeit interpreted through the gleanings of the collector of artifacts) paralleled another type of album as an inventory of objects prepared and executed through the filter of the camera.

Idiosyncrasies of individual human experience recorded through the lens disrupted the narratives of types or physiologies—popular beginning in early-nineteenth-century France—that attempted to subsume all human beings, all things of nature, into exceptions rather than rules. La Fontaine had intended this with his psychological typings. Sketches of urban characters, portraits and street scenes of Parisian life at the threshold of modernity, his physiologies were meant to codify images, institutions, and ideologies. Ray writes, "But while the new [photographic] technology seemed the ideal means of gathering the empirical data required by any system, almost immediately the first photographers noticed something going wrong." The relationship between name (language) and thing (object) could be rent asunder by what the chemical process on plates revealed. Ray continues: "what eluded classification—the distinguishing feature, the contingent detail . . . By showing not every Spaniard was not dark, every banker not dull, photographs effectively criticized all classification systems and ensured that any such system attempted in photography . . . would inevitably appear not as science but as art" (cited in Daston and Galison

297). Scientific evidence could provide guidelines along with variations; social evidence offered equal diversity within categories. As an early practitioner of collodion photography, a hazardous wet plate process flourishing between 1851 and the 1880s, Cajal was as interested in the science of the process as much as the art of the result. The transition from wet to dry plates, shortening the time between exposure and development, made the photographic process more convenient, even if the result of the wet plate was intriguing to him for its tones and light contrast. In addition, Cajal was obviously cognizant of the economic processes of modern times, lamenting somewhat the lost opportunities of perfecting the color process in a nation that had no infrastructure for exploiting it for profit. He was not directly involved in such entrepreneurial efforts anyway. Shortly after his marriage, he recalled finding in photography a compensation for having given up the brush and palette of watercolors for the rigors of the scientific laboratory, and he reveled in the chemistry of the collodion process, remarking about advances in the quality of photographic reproduction and his own missed opportunity to make money from perfecting a process heretofore of little use by photographers in Spain. Studying for exams kept Cajal from using his intellectual resources to compensate for the lack of financial remuneration at that point in his career, and finances were of the least interest to him unless produced by award or public acclaim.

The chemical development of photographic plates, so critical to the recording of detailed images and contrasts of light and dark, corresponded to his quest for a clarity of vision in the prepared slides of nerve cells. The contrastive details of color autochrome plates—a mosaic of tinted starch granules on a black-and-white base that allowed for the passage of light through color filters—attracted his interest as objects that unite theory and practice. He called this practical science *"unida a la abstracta o idealista, como el arroyo a su manantial"* ["indissolubly linked to abstract or ideal science, like the water of a stream and its source"] (*La psicología de los artistas*, 67). Cajal looked back at the early days of his attention to the chemical processes

> . . . llevé mi culto por el arte fotográfico hasta convertirme en fabricante de placas al *gelatino-bromuro*, y me pasaba las noches en un granero vaciando emulsiones sensibles, entre los rojos fulgores de la linterna y ante el asombro de la vecindad

curiosa, que me tomaba por duende o nigromántico.

Esta nueva ocupación, tan distante de mi devoción hacia la Anatomía, fue consecuencia de las insistentes demandas de los profesionales de la fotografía. Desconocíanse por aquella época en España las placas ultrarrápidas al gelatino bromuro fabricadas a la sazón por la casa Monckoven, y que costaban, por cierto, sumamente caras. Había yo leído en un libro moderno la fórmula de la emulsión argéntica sensible, y me propuse elaborarla para satisfacer mis aficiones a la fotografía instantánea, empresa inabordable con el engorroso proceder del *colodión húmedo*. Tuve la suerte de atinar pronto con las manipulaciones esenciales y aun de mejorar la fórmula de la emulsión; y mis afortunadas instantáneas de lances del toreo, y singularmente una, tomada del palco presidencial cuajado de hermosas señoritas (tratábase de cierta corrida de beneficencia, patrocinada y presidida por la aristocracia aragonesa), hicieron furor, corriendo por los estudios fotográficos y alborotando a los aficionados.

Mis placas rápidas gustaron tanto que muchos deseaban ensayarlas. Sin quererlo, pues, me vi obligado a fabricar emulsiones para los fotógrafos de dentro y fuera de la capital, instalando apresuradamente un obrador en el granero de mi casa y convirtiendo a mi mujer en ayudante. Si en aquella ocasión hubiera yo topado con un socio inteligente y en posesión de algún capital, habríase creado en España una industria importantísima y perfectamente viable. Porque, en mis probaturas, había dado yo, casualmente, con un proceder de emulsión más sensible que los conocidos hasta entonces, y por tanto, de facilísima defensa contra la inevitable concurrencia extranjera. (*Recuerdos* I, ch. XXVII)

[My cultivation of the art of photography was carried to the point of becoming a manufacturer of gelatine-bromide plates, and I spent my nights in a barn pouring sensitive emulsions under the red glow of a lantern and in the face of the wonder of curious neighbors, who took me for a goblin or a sorcerer. This new occupation, so different from my devotion to anatomy, was the result of insistent demands by professional photographers. Ultra-fast gelatine silver bromide plates, manufactured at that time to specification by the Monckhoven Firm, were certainly exceedingly expensive and almost unknown in Spain. I had read the formula for the sensitive silver emulsion in a recent book and I set out to manufacture it in order to satisfy my enthusiasm for instant photography, an unattainable activity with the cumbersome procedure of wet collodion plates. I had the good luck to hit the mark quickly with the fundamental procedures and even improve the formula for the emulsion; and my successful snapshots of action in the bullring, especially one taken from the presidential box full of beautiful young ladies (the occasion was a charity bullfight sponsored by and presided over by the aristocracy of Aragon), created an uproar, were sent around the photographic studios and excited many amateur photographers.

My plates pleased the people so much that many wanted to try them. Thus, without wishing it, I found myself manufacturing these emulsions for photographers in the capital and elsewhere, hurriedly installing a workroom in the barn of my house and turning my wife into my assistant. If I had been in touch with an intelligent partner who possessed some capital, an extremely important and perfectly viable industry would have been created in Spain. In these experiments, I had accidentally come across a method of preparing an emulsion more sensitive than those known till then and therefore very easily protected from the inevitable foreign competition]

It is evident from his own words that Cajal found in photography the confluence of three critical issues. First, the plates used to produce the image were scientific objects subject to the work of chemical emulsions on them. In his eyes, the preparation of the surface for the recording of an image satisfied his curiosity for experimentation and the perfection of a scientific process. Aware of his knowledge of this aspect of photography, Cajal took pride in having perfected the quality of the photo produced. Second, the art of photography also demanded a certain amount of time and dedication to a set of skills—those shared by professional photographers and amateurs alike—which, even if different from those of his anatomical sketches, set him to work long into the night to get them right. Third, the diffusion of Cajal's work with instantaneous shots (he even uses the word *snapshot*) created a reputation that he read as a potential market if he could write up the steps outlined in publications from abroad. Experimentation plus industry, with the labor of his wife added to the mix, produced the market conditions that modernity's entrepreneurs would rely on: more sensitive and better quality than foreign competition. Cajal was a man at the forefront of science, art, and (at least in theory if not in the pursuit of wealth) economics. The new social and cultural sensibilities—expressed through civic discourse—characterized modernity as "the ending of a traditional order . . . and the dawning of a new world of restless innovation" (Antonio 77) based on the rise of capitalism that fostered that technical innovation, so an optic of scientific development could very easily join with economics to promote new products, new knowledge, and new relationships between scientists, artists, and emerging values.

Chapter Two

The Curtain Rises on the Magic Theater of Life

Cajal, Master of Light and Color

Photography is actually a constellation of new technologies at the disposal of both artists and scientists beginning in the nineteenth century, especially over the latter part of the decade of the 1830s.[1] Olivier Debroise reiterates the fact that photography is not one but a number of innovations associated with inventors of various types: scientists, graphic artists, painters, chemists, and opticians. The convergence of a number of advancements in several fields brought together the circumstances necessary for a more permanent recording of images, not just the fleeting ghosts of early photographs. As Debroise concludes, it is not only the production of the image that was revolutionary but its preservation: "At the beginning of the nineteenth century, modern chemistry and optics made it theoretically possible to retain images . . . The invention of photography was less a question of finding a way to reproduce images than of retaining them, fixing them permanently on a support" (18). The Daguerreotype ushered in the material proof of a chemistry of photography, and a method of recording that allowed for a return to study the image.[2] Therefore, a case for photographic images as scientific artifacts coincides with them as aesthetic objects. Both cameras and the photographs they produced collected evidence of a modern sensitivity to the reproduction of images of the world, a world increasingly in transition and increasingly less attached to "a privileged viewpoint at a particular moment" (Everdell 11) that had driven observation previously. Cajal's laboratory work in histology, and his experiments with photographic chemicals, color, and light bridged the nineteenth century's belief in the ability to understand the whole and new, disruptive ways of looking at nature and culture.

Bertrand Lavédrine, in a recent publication on optimal ways of preserving photographic images, establishes the critical environment for the invention of photography as an atmosphere that supposes scientific processes at the service of the development of such an image—be it scientific or artistic. He summarizes that "The era of the first photographic processes was one in which the knowledge of chemistry and physics was developed enough to allow a determined inventor to make a permanent image on a photosensitive material through the effects of light" (22). The negatives and their positive prints are documents filled with information about the state of culture, science, the arts, economics, and the family at specific times in history. These are recorded and framed—with increasingly sophisticated techniques—by observers of the human landscape as perceptions change alongside social transformations. A relatively recent addition to capturing images, photographic processes advanced rapidly, with innovations flooding the field and, with time, "processes that occupied, and even dominated, the field in the past . . . [even] disappeared entirely" (Lavédrine 5). There was always the desire for a new and better product. From collecting specimens to document, a dedicated photographer added expertise in chemistry to his repertoire as a nod to the modern.

Max Kozloff follows many of the theories of Walter Benjamin on the potential for scrutiny of photographic images in search of traces of the new processes as well as "a subliminal expression they could have offered or betrayed" (7). Readings of photography as an intellectual activity pursued by scientist and artist alike take place in the realm of the collection, the album, the exhibition. While the spark of time at the moment has been captured, Benjamin finds that the space of perception has changed. Kozloff, like both Benjamin and, subsequently, Roland Barthes, recapitulates the argument that "the camera may immobilize its subjects, yet it by no means petrifies them" (7). The decay of the past—the faded snapshot, the shadowy tintype on metal or darkened ambrotype on glass—appears as a ruin outside chronology, whether in a collection or among the objects inherited from the past as a collage of objects. The image is, in Walter Benjamin's view, extracted from its context and freed from a singular reading or meaning, cast into the field of observation of every generation of spectators, and given an afterlife of consumption and multiple interpretations. Through the science of photography as the chemical fixing of moments that have happened already, we are reminded of the past through associations with images emancipated from their original contexts and cast into new encounters with viewers and their experiences. Unknown people and places become analogues of previous times and spaces, not a linear recovery *of* their time and space.

Long a painterly tradition, the still life photograph of the eighteenth and nineteenth centuries captured a variety of objects at once. In a scene composed by the photographer using a variety of objects from landscape to sculpture, from human beings to domestic pets, from familiar flora to uncanny fauna, and from the objects of the laboratory to the volumes contained in a personal library, traces of various worlds unite and collide. The time-consuming, and of course subjective, act of painting was cut short by the ever-evolving processes of photography, allowing that, in the words of William Henry Fox Talbot "'the whole cabinet of a Virtuoso' could be represented in a little more time than it would take him to make a written inventory describing it in the usual way" (cited in Martineau 7). From adorning scientist's catalogs, cabinets of curiosities of scientists and explorers, the photograph moved on to allow the collecting of items in a new, more modern, way that occupied less space and evoked new relationships.

As seen in fig. 2.1—*Bodegón, método interferencial de Lippmann* [*Still Life—Lippmann Interferential Method*], one of Cajal's color photographs illustrates an example of the Spanish *bodegón* or still life genre, captured with the use of a glass plate and fine grains to achieve maximum definition. In addition to the more usual flora of the region, such as plants and blossoms, there are fruits, drink, and other accessories. The appearance of a particular item is unexpected but telling: the microscope. The two female statuettes—indicative of a Victorian-era aesthetics much like those captured by Hippolyte Bayard or Louis-Rémy Robert in the mid-nineteenth century—balance the scene in harmonious display. They also frame the objects on each side just as the flowered cloth emerges above and below to contain the still life. The space of the composition is enclosed by a tight focus on the objects, leaving no light or outside intrusion on the display. It is rendered time-less—neither day nor night, and no sense of season—but also displaced since it floats freely in its own space. To the left of the staged scene, a young maid carries a ceramic jar of some sort, while to the right a semiclad female figure evokes the Venus de Milo with the absence of

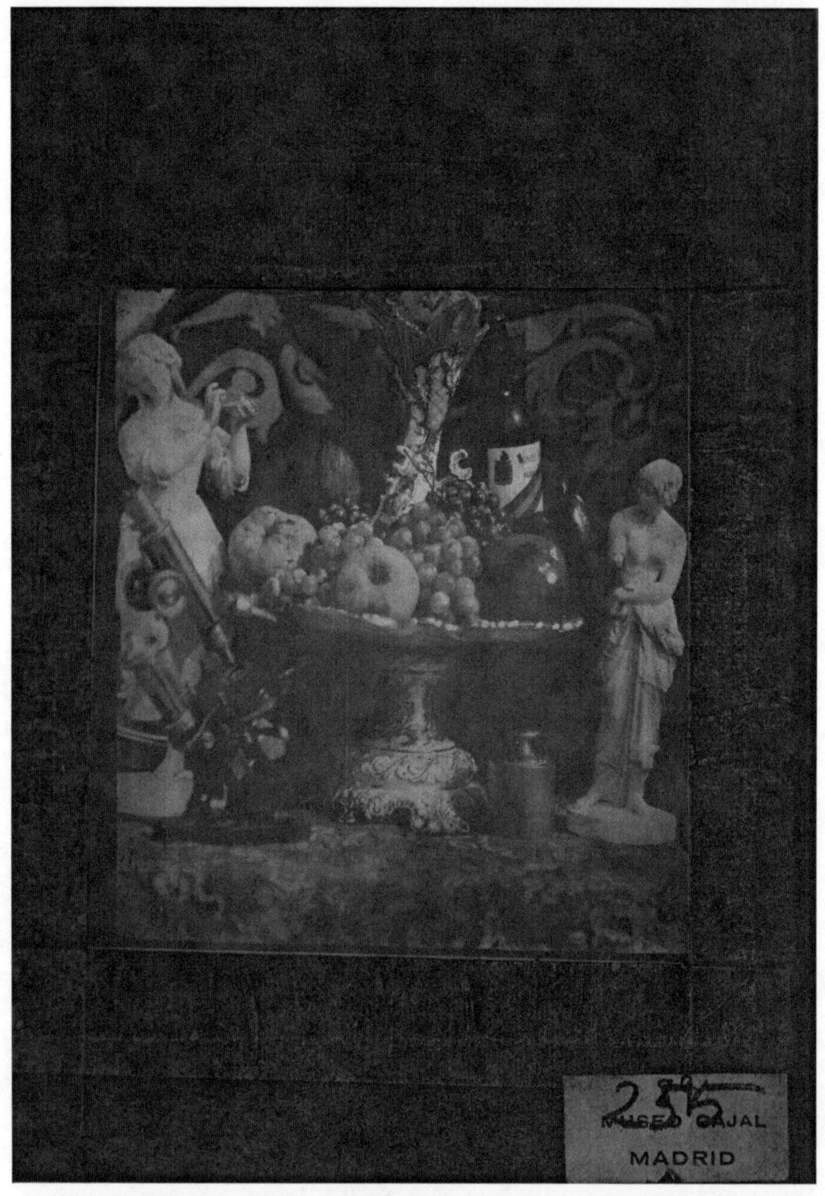

Figure 2.1. *Bodegón—método interferencial de Lippmann* [*Still Life—Lippmann Interferential Method*]. Santiago Ramón y Cajal. Legado Cajal (CSIC). Instituto Cajal. Madrid, Spain. Used by permission.

a right arm. Classical in proportion, both statuettes carry with them the feeling of a traditional aesthetics: they embody the beauty of the human form in the work of art.

Grapes and other fruits of an abundant harvest occupy center stage on an ornate dish, complementing the bottle of wine in the background and the flowing, vinelike patterns of the draped cloth. A small brass weight in the right foreground glints in the light, and that object of technology used for printing or perhaps for a counterweight in scientific experiments finds equilibrium in the bright metal glint of the microscope at left. There is a totality here, the representation of completeness, with a seamless and harmonious reproduction of nature in the still life of fruit and foliage, mixing the natural and the manu-factured, the observed and the technology of observation. The female figurines do not disrupt the flow of abundant nature, nor does the machine included among the elements of nature seem out of place.

The presence of the microscope in the foreground at left can be read as a stand-in for the camera used to record the image. Its eyepiece and the lens function similarly to allow access to those "enigmas" and other mysteries found within the natural world, if only devices could be invented to pierce what he had called that "dark cloud" of nature's unknown aspects. The microscope also preserves that material object at the cutting edge of science that had become part of his natural world as it had taken over Spanish mass culture. Cajal's fascination with the theory of the latent image provides a clue to his inclusion of the microscope here: objects, details, and outlines appear seemingly out of nowhere and from nothing when such inventions create the opportunity for new sight. Both make evident the technologies that enhance vision; they then demand new ways of looking to accompany the new means of seeing. While the notion of looking through the lens may cloud the presence of that device itself—the intermediary disappearing as the eye connects with the objects—the microscope demands that the viewer consider not just what is seen but *how*—by what means—it was rendered capable of being seen. The metallic finish of this manufactured aid to sight gleams in the light as a trophy of scientific research, worthy of its placement in the foreground. Since it was a gift to Cajal in payment for his work on helping to respond to the cholera epidemic of 1885 in Valencia, the Zeiss microscope brought the composition of the still life back to life as a tool for restoring the health of the inhabitants. With the microscope, the scientist could see the cause of death and respond with a remedy to make the technology become an aid to life. Not merely "a mechanical device used for making copies of the natural world" (Martineau 7), the camera that captured the objects in this frame was a conjunction of physics, chemistry, and art that produced images chosen, framed, and focused by the photographer. They are also images that are repeatable and endure.

For Cajal, there was always a challenge to be found through the lens, in a still life or otherwise. His montage of objects—vases, flowers, in another print even one of his daughters in a flowery kimono—includes all the elements surrounding him. The innovation of color photography gave Cajal the power to test the lens and the processes by introducing pattern, texture, depth, and shades of color. His use of a combination of registers in his choice of objects rewards the observer who may feel two shocks: the interjection of modern innovations in a traditional genre, and the intense colorization of the articles in the photograph that reproduces the colors of nature. Returning to Prodger's remarks once again, for the scientist photographs became an illustration of a person, a scene, a collection of goods and objects, but they were "experiments in their own rights" (xxiii). The data they provided the eye of the scientist confirmed or contradicted a hypothesis, revealed the validity or disclosed the inaccuracy of a proposition, or represented a situation at a certain moment in time, in addition to exemplifying the power of new techniques. Concerned with both, Cajal kept a record of his specimens—cells, family members, biological structures, Spanish landscapes—that embodied a fixed record of the development and evolution of the very medium that had preserved it. The fact that an album or collection of photographs could be examined at leisure, that they represented discrete moments in time that were noted (written) on the images themselves, they could help unravel complex events, structures, or processes that otherwise might prove difficult to analyze. A dendritic tree or the complex changes in a cityscape could be accessed through distinct images to arrive at a possible synthesis of its elements.

Obsolescence seemed to accompany all aspects of modernity's paradoxical "tradition of the new" (Antonio 77) as cultures adopted values that replaced modes of the past, and new instruments for old. Cajal demanded in his exordium to future generations of photographers in Spain that science had to find a central place in subsequent generations of Spaniards. He concludes that *"los que vamos para viejos . . ."* ["those of us on the threshold of old age . . ."] (*Fotografía de los colores*, 18) leave a legacy to be taken up and challenged with procedures and techniques not even dreamed about yet. What he calls a "sad reflection" (141), was that he felt he was born too soon to see where the marvels of photography might lead. Daston and Galison summarize that the "family of techniques" comprising photography (125), one that included instruments of scientific discovery as well as the imaginative production of aesthetic images, cannot be underestimated for their potential to increase knowledge. Crossovers between cameras and microscopes encompassed the potential of each, spurred by the knowledge of lenses and light inherent to both. Ramon y Cajal's critical and innovative work in the fields of histology and microscopy, on the one hand, and black-and-white photography, then later, color photography, on the other hand, illustrated some of the most important shifting paradigms of the second half of the nineteenth century that focused on the cultural value of seeing as signifying access to the authenticity of the social and natural worlds. As radical changes in social life occurred—albeit in fits and starts in Spain over the nineteenth century—there was an immense desire to explore all modes of visuality to access these in significantly different, nominally modern, ways. The focus on the eye included military uses, the sitting room and studio, medicine, the reproduction of works of art, recording of historical monuments, and scientific investigation. With this passionate desire for new ways of seeing, Hope Kingsley finds that optical technologies coalesced simultaneously around both art and science: "Optical aids predated the photographic camera, though not by much; problems with focus and lens aberrations meant that they were actually practicable for less than a century before photography's invention . . . photographic cameras incorporated earlier designs and corresponding representational systems" (78). The focus on the power of sight did not just appear with the camera lens, but with the microscopic aids, telescopes, and corrective lenses of earlier eras. With such a vast optical grid, one might ask whether the natural world could be represented adequately by the camera and subsequent photographic product, or how reliable the product of these technologies is. Is the image a transcription of the visible world, or does it have the same limitations of the human eye (partiality, field of vision, luminescence, etc.)? While Spanish cinema as a launching of new technology that trained consumers in the culture of the eye has been the subject of many more critical studies than the photographic work of the same era, the black-and-white as well as color photography of Cajal has been a fairly overlooked source for the implementation of *other* aspects of visual culture in Spain between the 1830s and the 1930s.

As Cajal wrote in his volume on color photography and on his relationship with "*la placa sensible*" ["the photographic plate"], this family of techniques developed at a rapid pace and acquired increasingly sophisticated capacities to reproduce the natural world and its inhabitants. As Prodger notes with exactitude, in Cajal's time "photographers had to be chemists, craftsmen, and time managers as they shuttled from darkroom to studio and back" (7). After careful collection of data or objects, the photographer had to decide how to capture the image, light it, focus, and then develop the result. Mastery of the processes as well as the setup was critical. Cajal's own burgeoning cultivation of the photographic process as technology, as well as the photograph as capable of capturing the world's "bellezas naturales" ["natural beauty"] for the curious observer, emerged in his conclusion that this invention was capable of getting closer and closer to reproducing the special attributes of the human eye. From dissecting the eye, to sketching its structures and tracing its functions through the brain, then to mimicking its production of images, he linked biology with technology and art. Reviewing Cajal's lifetime of what he termed the distractions, pleasures, and even "poetry" (*Fotografía de los colores*, 15) of the aficionado of photography from a youthful (and "innocent") enthusiasm for daguerreotypes, through an adolescent romance with the smell of the collodion plates, to the gelatin bromide of Bennet and Monckhoven, Cajal closed his discussion with a reference to the power of photography as a compensation for the fading of sight associated with old age. Toward the end of his introduction to the 1912 volume on color processes, he celebrated and lamented this in the same breath:

> Dicha grande . . . ha sido el poder asistir a la *éclosion* del procedimiento autocrómico y saborear sus encantos, antes de que la terrible *catarata senil*, empañando nuestro objetivo ocular, baje el telón sobre el mágico teatro de la vida. No concibo tormento mayor para un admirador de la naturaleza que este cruel destierro de la luz, decretado por la senectud confabulada con la enfermedad.

> [What an enormous pleasure . . . it has been to be able to be part of the launching and popularity of the Lumière

autochrome process and to savor its graces, all before the
terrible cataracts of old age, fogging our ocular lens, lowers
the curtain on the magic theater of life. I cannot conceive
any greater torment for an admirer of nature than this cruel
exile from light, decreed by the process of aging and by
disease."] (*Fotografía de los colores*, 16–17)

Whether light is exiled from his universe, or he is exiled from the
light, Cajal evokes the human decay of aging with the power of vision
to connect human beings to their environment. Losing the capability
of observation would cancel his access to the world.

Using the rhetorical device of a question posed to readers of the
manual, perhaps members of a younger generation that would take
his work further, Cajal both personalized the taking and collecting of
photographs and affirmed a desire for the penetration of the process
and its products into all aspects of Spanish culture. It is from this
passage that the title of this chapter is taken. The vibrant theater of
life disappears from view as the fog of the cataract lowers across the
human eye to rend the world and the human subject's perception in
two. Before the possible loss of sight, with the subsequent turning of
the microscope into a useless tool for the former scientist, Cajal made
certain that he recounted the historical trajectory from his earliest days
to 1912, and from simpler black-and-white processes to plate, mirror,
glass, Lumière autochromes that could be projected or viewed through
a special apparatus (the chromodiascope), and paper prints in the early
1900s. The manual is a testament to Cajal's intellectual dedication, as
well as evidence of the breadth of his knowledge and interests.

For Cajal, the details were as important as the finished product,
with the effects of shadow, tone, light, and color at the top of his list
of priorities for a quality photograph. If technology were to aid the
eye, it would have to be equal to or better than that organ. Its ability
to record motion unseen by the human eye was the starting point of
that characteristic quality that would capture time in new, scientifically
accurate ways.

Ya en plena madurez, saludé regocijado la aparición del
Autocromatismo de Vögel y la exquisite sensibilidad de
las emulsiones argénticas. La placa pancromática actual se
identifica en sensibilidad cromática con nuestro ojo. Ya no

traduce solamente el rayo azul como debe ocurrirle al pez de los abismos del mar; impresiónase también, en determinadas condiciones,en presencia del verde, el amarillo y el rojo. Gracias al admirable invento de Vögel hánse aclarado las mejillas y las rosas, y se han obscurecido como debían el cielo y las violetas.

[Once I reached my adult years, I enthusiastically celebrated the appearance of the three-color sensitivity process refined by Hermann Vögel (in the 1870s and 1880s) along with an exquisite sensitivity of silver emulsions. Today the three-color or four-color plate is as sensitive to color as the human eye. Now not only rays of blue appear on the plates as fish at the bottom of the ocean might perceive them, it is impressive that we can also see, in specific conditions, shades of green, yellow, and red. Thanks to Vögel's admirable invention, the blush of life in cheeks and roses has been lightened, and the darkness of the sky and of violets have been toned down as they should be.] (cited in Romero 1)

The "right seeing" of objectivity in the observation and reproduction of nature is evidenced in the last phrase related to the color of flowers "as they should be" seen, not imagined, without documentation or reproducible images. From his younger days as a so-called enigmatic and malevolent figure working under the red glow of a lantern in the barn (the reference appears in his own memoir), Cajal progressed to more modern techniques. These included various types of color processing and more scientific cross-commentaries based on his studies in the comparative anatomy of the eye. As Rodolfo R. Llinás argues about Cajal's work in anatomy and neuroscience, a principal theme running through the histologist's experiments and observations is "his realization that understanding brain function must be more than a piecemeal endeavor. To him, such understanding came about only with reference to 'the big picture'" (77). That big picture included general observations on photography and not merely individual images, as well as a depiction of the brain, from schematic drawings to detailed empirical data. Cajal compiled information on the structure of the human eye and their echo in the sensitivity of the color plate. He included psychological pathologies as well as the systematic function of healthy structures. The structure and function of the eye itself was an object of scientific study for Cajal as much as how it worked and what it—or its surrogate, the lens—registered.

The emergence of new technologies produced by Germany's Zeiss Optics (e.g., the four-lens Tessar camera), compound lens microscopes, the French Verascope stereoscopic and pocket camera, the Steinheil Antiplane Aplanat lens, fast-developing silver bromide emulsions, and the like accompanied a search for increasingly accurate knowledge about the world in the form of visual access to concrete details of the whole picture made visible. These details could then be made legible to others by means of codifying the images in collections, albums, compendia, and atlases. The scientist systematically gathered in an atlas the objective traces of the natural world to make sense of that world, name its components, and in the process choosing to record some images but omit others. The repetition of objects included in the atlas by the trained practitioner emphasized the process of gathering them, as it simultaneously intended to reveal "how things look" as a collective whole. Such "dictionaries of the sciences of the eye" (Daston and Galison 22) were repositories that preserved and stored the forms of the world as understood by a culture in a given time period. If the moment happened to be turn-of-the-century Spain, then an understanding of the modern as a culture of visuality appeared in the value accorded scientific cabinets of images, albums of photographic proof, photographic records, personal albums, and portraiture. For Benjamin, "the beginnings of photography in portraiture mark . . . a transition from cult to exhibition value. . . . Early portraits are, as a consequence, auratic, a property, which is dissolved as photography moves from evoking remembrance to bearing witness" (Caygill 107). From re-creating the past, the portrait had been turned to record the present. Portraits have an intrinsic tension, the "hypothetical, sometimes tense conduct, worked out by agents in an unstable process" (Kozloff 8). Storylines can be mingled, the familiar with the uncanny, the staged with the candid, the personal with the public, the close up with the panoramic shot, the evident with the mysterious or the monstrous. Then the processing of the image adds the dimensions of light and dark, contrast and fade, color and tone, highlight and low light.

How might we read the cultural values of the nineteenth century, or for that matter of the early twentieth, on the evidence of photography? Benjamin delved into incongruity and uncertainty as the contributions of these technologies to the modern fabrication and perception of the image. Thus, the photograph as an event requires careful analysis. Cajal moved from early portraits to later self-portraits and architectural, geological, and familial portraits. It holds true that the modern observer may not identify with the stance, the look, or the pose of the individual(s) captured in each phase: "we [might] consider nineteenth-century manners as 'heavier' and more aloof than ours. But their self-honorific stance was naturalized by their culture and a community of values agreed on by observer and observed alike. . . . They would have looked far stranger, at the time, had they been relaxed. Everyone was stiffened . . . by ideas of appropriate conduct" (Kozloff 9). "Right" conduct, like "right" vision must be placed in context. Which subjects were considered to be in the realm of propriety, what created the equilibrium of the portrait, what of the length of the take, are aspects that can appear puzzling. They are emblems of otherness to us but not so in terms of the culture of their time. But in that very difference from what occurred after the shot was taken, from the roads taken by cultures since the photograph was made, lies a century of clues that beg us to negotiate Cajal's work in terms of marking a transition and not mere preservation or monumentalizing the objects of the past.

Cajal's children and his wife were the most frequently portrayed subjects, sometimes with himself included in the scene, other times not. Yet there appears to be a difference in the portraiture between the two types of image. In self-portraits, Cajal seemed to relish the notion of the human face as a countenance adjudicated by the scientist occupying that body. Thus, his glance when posed alone was rarely toward the camera but more remote, pensive, distant, high-angled rather than downcast. His role was to think. When taken of others, Cajal's photographs reverted less to the intellect behind the countenance than to the person as a backdrop, a human element occupying a space alongside a landmark or natural space. These figures were frequently his sisters or brother, or his wife, as a marker or a trace of an instant that can be witnessed through their faces. His countenance had to look different from theirs as his profession dictated objectivity, intellectual absorption, and a dedicated will, not pleasure. Their faces were domestic, familiar, and similar to others one might find in the street. The camera could be used to mark the singularity of this man of science in several ways: first, his distinctive facial features and second, his dexterity with the photographic apparatus to capture the images. If, as Cynthia Freeland writes, "the portrait artist is an alchemist who seeks to make inert physical material 'live' and show us a person, an actual individual whose physical embodiment reveals psychological awareness, consciousness, and an inner emotional life" (1), it follows that this task sounds tremendously mysterious, on the one hand, and just as tremendously powerful, on the other hand.

The inner life of Cajal as a scientist had to be supported by visible and material accoutrements of the intellect in order to give evidence to the eyes of others what the mind of the scientist was pursuing. On the surface, the man of science looked quite like any other father, brother, husband, or gentleman of a certain social class of the times. The corduroy or pinstriped suits, buttoned vests, watch chains, dark laced shoes, beard, and look of seriousness can be found on men of many professions. Both a gentleman and a scientist, Cajal's self-portraits conflated the two worlds. Yet there had to be a record made of something that distinguished Cajal and his peers from the rest of the crowd. The frames of his self-portraits were filled with equipment, books, the worn and dirtied surface of the workspace, piles of papers, and a look of quiet determination. When he placed himself at the center of a family portrait—as in fig. 2.2—*Familia* [*Family*]—Cajal stared straight ahead at the camera, in control of the shot and of the position of those included in the frame. His arms encircle two of the four children posed alongside him, and they form a quiet family group surrounding the patriarch. For the observer there is no doubt about who was in control. The inner life of his family was less visible or, perhaps owing to the invisibility, entirely absent or inconsequential, except for the few minutes they spent waiting for the image to be taken. That life had to be speculated about. Especially true for an audience whose general knowledge of things scientific was insignificant, the material traces of science represent that mystery in facsimile, as the technologies of the laboratory are placed in front of us to surround the man of intellect. We must make the leap to reconnect the Zeiss microscope with Cajal's Nobel Prize, the microtome with his slide preparations, the vials and flasks with his stains and slides as indicators of his type of life and how he spent his time (aside from taking the photographs, at least). That he returned to family portraits time and time again is evidenced by the number of images that remain in the collection, as well as by the naturalness with which the children sit

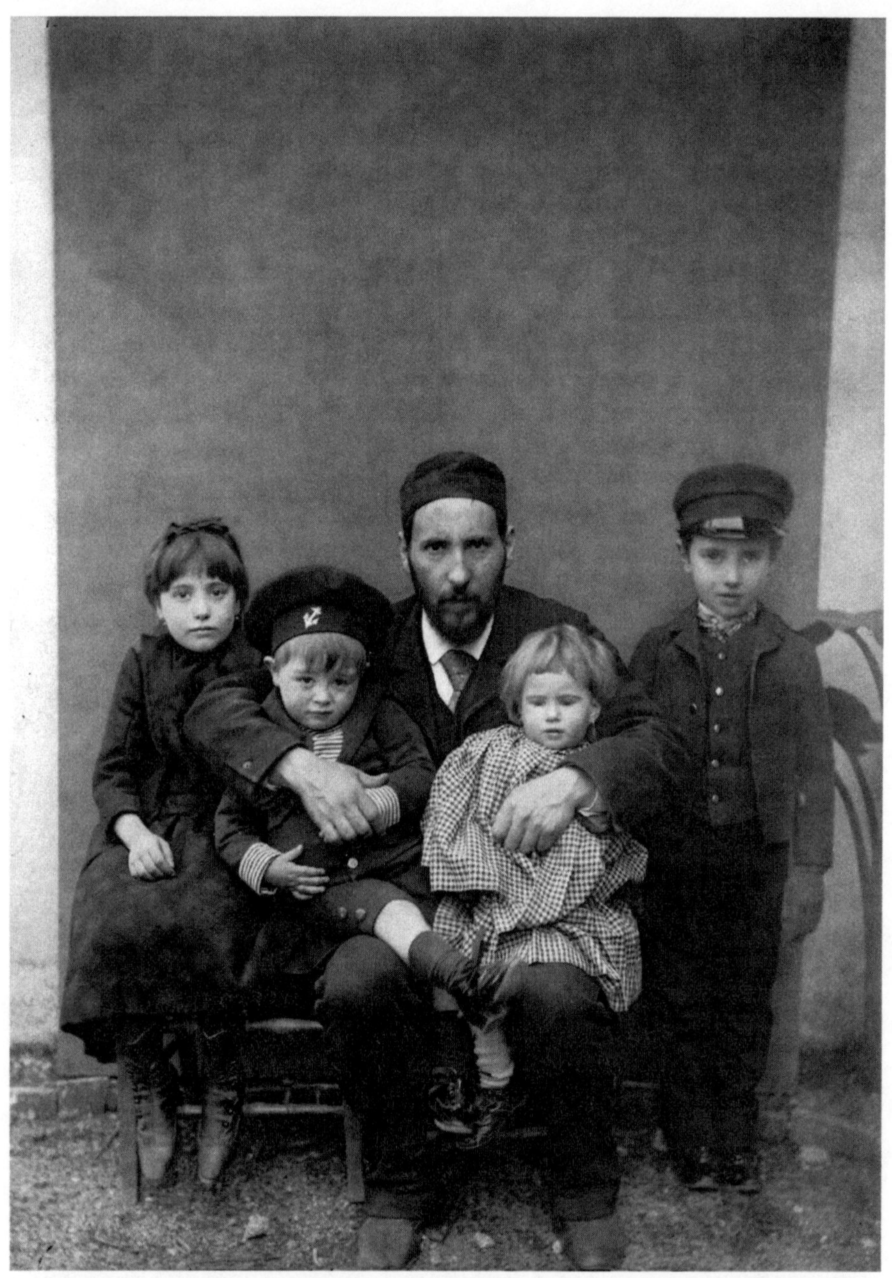

Figure 2.2. *Familia* [*Family*]. Santiago Ramón y Cajal. Legado Cajal (CSIC). Instituto Cajal. Madrid, Spain. Used by permission.

for the shot. They are used to posing. Here, the gravel path and the stucco wall indicate a scene outdoors, as does the clothing keeping them warm as they wait.

The complementarity of an emotional life to his intellectual one in his portraiture of the family manifested the control of the scientist over his environment—the poses, the arrangement of people, the setting of the lens, the click of the shutter. One was left to discern the tiny indicators of the lives of his children outside the frame of the photo, hinted at by a look in their eyes or an inkling of the desire to run off. Cajal as photographer—and, after all, their father—captured them as "real" girls and boys with both inner psychological characters and outer characteristics, and as *his* boys and girls, members of his family who had names and were individuals who inhabited real space, and grew and changed over time. His documentation of their external appearance as a group filled albums with evidence of time's passage, presence and absence, and with his own relationship to them and among them. Our knowledge of Cajal through his images parallels his self-knowledge through them, as well as the relationships he found between the discoveries of the laboratory and the life cycle of his family. How he presented himself looking at them through the lens also displays for us what he was doing (and thinking) as he captured what they were doing. What they might have been thinking is the absent piece of the now-lost moment, but something beyond routine can be inferred.

Cajal's family portraits toward the end of the 1890s multiplied. He had been at work on methods and processes of development for several decades, and the clarity and definition of the faces on this black-and-white image are impressive even as they show little emotion. In a panoramic family portrait, experimental photography and familial documentation come together. Fig. 2.3—*Familia* [*Family*] shows the children and their mother posed against a wall, from youngest to oldest, with the exception of a daughter who is taller than her mother. The steps and stairs of this dignified group reflect the composure of a family that can be put on display by the photographer, who also happens to be their father and husband, respectively, and a rising scientist in the public eye, without any qualms. As proud as he was of his photography, he must have been equally sure that the subjects would measure up. Hands folded, or at their sides, the young men

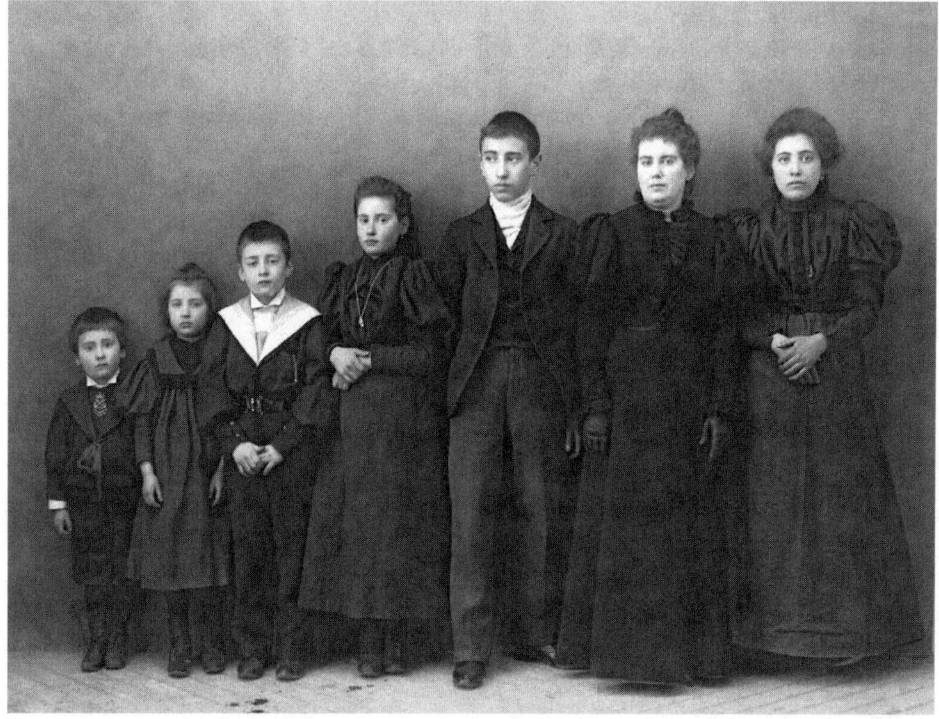

Figure 2.3. *Familia* [*Family*]. Santiago Ramón y Cajal. Legado Cajal (CSIC). Instituto Cajal. Madrid, Spain. Used by permission.

and women of the Cajal-Fañanás family are not about to disturb the session. They obviously have dressed in their finest clothes for this more formal than usual shot. They stand quietly, the youngest to the oldest displaying a calmness and willingness to be part of a family portrait. It does not seem to be an experimental shot with multiple versions or differences in lighting or texture, but rather a formal composed record of the group. But this image is not a monument to stasis either, as the numerous photographs of them across their lifetimes show change, decay, growth, and disappearance (death). As the children mature, the spark of a moment in the past is recaptured in the photo collection that documented the comings and goings of the members of the family, as it simultaneously recorded changes in photographic quality. In this case, all are present as a document of a moment in time when they shared this event that included the man behind the camera. Science is part and parcel of their home life.

In his study of how and why photography is "unclassifiable" and what its "disorder" might be, Roland Barthes asserts that the singular photograph is always bound to its referent and that the observer, in order to gain some knowledge from an encounter with the image, "requires a secondary action of knowledge or of reflection" (*Camera Lucida* 4–5) to bring it to life. As one of the many objects or "things" belonging to the world, the photograph "mechanically repeats what could never be repeated essentially. In the Photograph, the event is never transcended for the sake of something else: the Photograph always leads the corpus I need back to the body I see; it is the absolute Particular, the sovereign Contingency" (*Camera Lucida* 4). So the photograph is not literally transformed but it transforms and animates the relationship between viewer and object, with the personal needs of the observer encountering the recorded image. The modern reading of the image is contingent on a "disorder" or disruption of the ease with which images might be read. The dialectic between what is desired and what is seen takes place across the now-closing distance between the taking of the photo and its being torn from that moment in which it was embedded. Alongside Cajal's commentary on color photography and his remarks on the fatal illness of his son Santiago, a fact uniting the social body and the biological body, the photographs of the family acquired a sense of loss for him. Santiago was there one moment and gone the next, strangely akin to the loss of coherence and linearity associated with modern times. Barthes continues: "In front of the lens, I am at the same time: the one I think I am, the one I want others to think I am, the one the photographer thinks I am, and the one he makes use of to exhibit his art" (*Camera Lucida* 13). Any notion of the "authentic," then, disappears with the uncomfortable confrontation of oneself by oneself, oneself by others, in a curious and changing mixture of familiar and defamiliarized encounters. After the moment of observation, "when the photograph is no longer in front of me and I think back on it" (Barthes, *Camera Lucida* 53), when the album cover is closed over the collection of images, a spark of perhaps unnamable emotion can be triggered. Santiago's illness and death are one example of the photograph's power for Cajal; even after the image is filed away his son is the object of his recollection in his memoirs.

There is a plethora of evidence for concluding that Cajal exhibited exemplary dedication to methodical gathering, labeling, archiving, and collecting, in general, attempting to make sense of the world around him without missing a detail. His rapturous remarks on photographic albums are only one piece of evidence related to methods of comprehending the world as completely as possible. Examples ranged from the carefully written paper labels for his files and specimen boxes, to the columns and lists of course material in preparation for examinations, from his ordered notes of German vocabulary to the multidrawer cabinet with histological preparations that he carried with him on scientific travels, and from his notebooks of sketches and drawings to the piles upon piles of repeated portraits and self-portraits. These appear in stereoscope, with and without retouching, and with greater or lesser detail. There does not appear to have been an aspect of his inner or outer life that Cajal did not document multiple times, or preserve with each new technology available. Inheriting the power of Daguerre's revolution, Cajal became fascinated with the camera's potential to be a critical part of modern Spain, whether during picnics in Zaragoza or as an officer in Cuba in 1874. He was recorded at home looking quizzically at the human skeleton he assembled as a medical student; catching the harvest of orange pickers in Valencia, or at meetings of his gastronomy club; posing with full academic regalia for official portraits; and capturing his children with toys, without toys, smiling and sullen, skittish and still. The inner life of his subjects is less transparent.

His own collections of photographs are artifacts of the science of collecting itself, as albums of faces and places brought together by an individual for study and observation over time. In one entry in his recollections, he asked the reader to ponder "¿Habéis pensado alguna vez en lo que significa un álbum de fotografías?" ["Have you ever stopped to think just what a photograph album means?"] (*Fotografía de los colores* 17). He followed up on this question with a disquisition on the symbolic reversibility of time by means of images frozen in other moments but resuscitated anew, without the ravages of sickness, childbearing, or age, recapturing the spark of youth in a flash. All is done through the contemplation of the photographic image. Although obviously not a very flattering portrait in words, Cajal used his wife as his first example, referring to her now as "la robusta matrona rodeada de retoños: desdibujada por la grasa, convertido el artístico jarrón en imponente cuenco. . . . Por fortuna, ahí está vuestro álbum. Miradla rehabilitada y transformada en grácil doncella" ["the robust matronly figure surrounded by her offspring: a lump of fat, the artistic pitcher become a squat, imposing jug. . . . Thankfully, you can turn to your album. Just take a look at her transformed back into her old self, a delicate young girl"] (*Fotografía de los colores* 18). He added to this montage of images that included Silveria (see fig. 2.4—*Retrato: Silveria, Jorge, Pilar*) a comparison of the figure of a veteran, now slow, lame, and overweight but in the proverbial album a recovered heroic icon of glory on horseback bedecked in medals from his service to the nation. Cajal's conclusion is that the album is an ongoing process, a collection that is never closed and complete but that it evolves: "puede todavía enriquecerse, y a la vieja colección de fotografías en negro aãndiremos la nueva serie de fotografías en color. Y todas estas pruebas tendrán derecho a nuestro entusiasmo" ["(It) can still become enriched, and to the old collection of black-and-white photographs we add the new series of images in color"] (*Fotografía de los colores* 18). Scientific technology advances as time passes, the photo captures the transitions of both. Neither present nor past exists without the other, and time does not stand still but moves in fits and starts in a dialogue between youth and age. In several photographs, Silveria appears in the living room or on the porch or balcony of a home, accompanied by her young children. She is often dressed in everyday clothing—sweaters, leggings, and a long skirt—attesting to a pause in her activities that she has taken at the behest of the photographer. She frequently looks down modestly, as if preoccupied with keeping her infants under control but aware of the fact that she will resume her work after this moment has ended. Or maybe she is lost in the thoughts of some other place and time, or in the difficulty of holding the pose for the time it takes to register the image. These are not the formal poses of studio portraits but of stilled domesticity, the inner life of the subject not made clear from her visible expression. The potentialities that Walter Benjamin evoked in the contemplation of photographs, even early portraiture that retained the possibility of lapsing back into the auratic, appeared in Cajal's assessment of the technological medium as a connection between the material person standing in front of him and the woman snapped earlier on an excursion in the country. Caygill discusses this transitional property of photography as a move from a one-on-one recognition to juxtaposition, decontextualization, and an interjection of

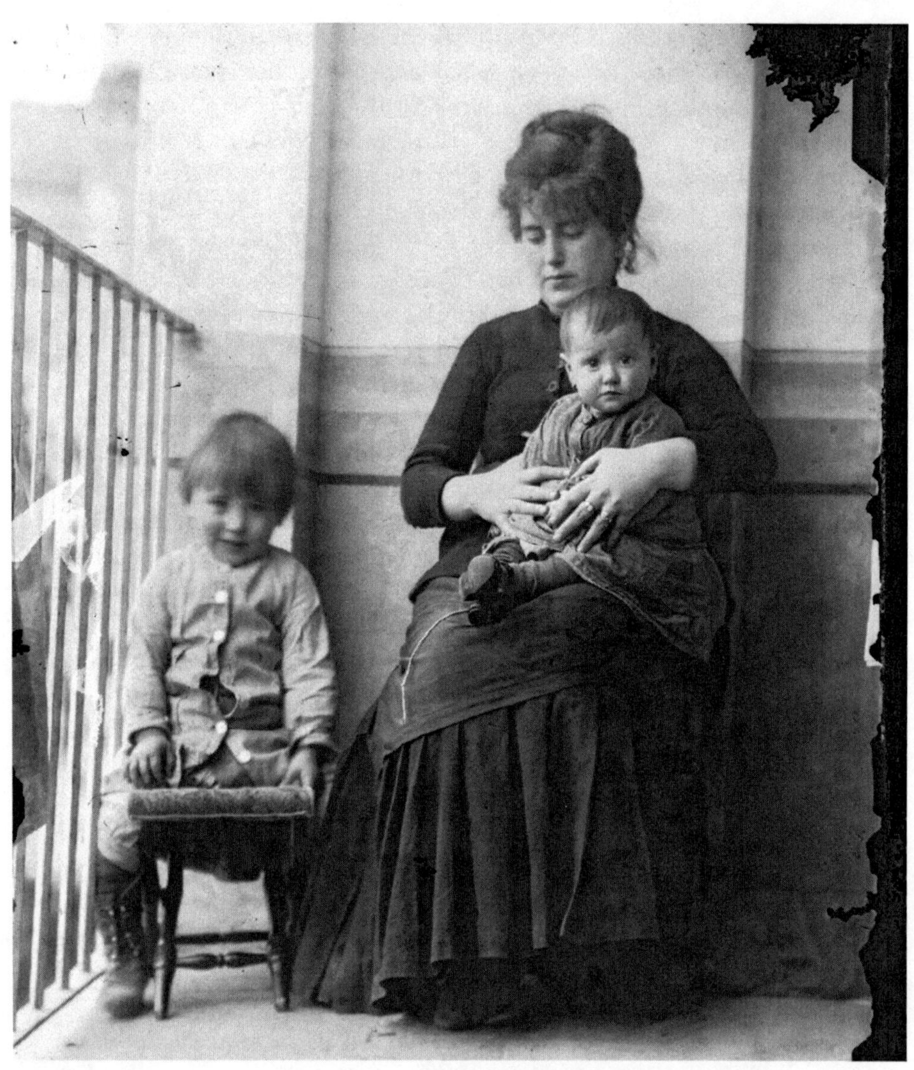

Figure 2.4. *Retrato: Silveria, Jorge, Pilar* [*Portrait: Silveria, Jorge, Pilar*]. Santiago Ramón y Cajal. Legado Cajal (CSIC). Instituto Cajal. Madrid, Spain. Used by permission.

the unknown or the uncertain, through the discrepancy among images that challenges the viewer (107). One may not recognize an event or a face, but evidence of a historical moment is there to describe with new eyes as it disturbs easy reconstruction. Like Cajal's union of fragments in science, the photograph demands the bigger picture.

As was previously mentioned in chapter 1, the simple collection of birds' eggs and nests from the woods around his childhood home was Cajal's earliest experience with taxonomic collection and the "dictionaries of the sciences of the eye." These were assembled in then-unknowing anticipation of what he later did with drawers full of histological slides—only recently carefully cataloged by García-López, García-Marín, and Freire—and with albums of photographs that document the inhabitants and vistas of Spain, Europe, and America. The careful labels he penned for them respond to a desire for ordering the evidence; they are legible even from afar and preserve clearly his method of inquiry. His preference for "*la recopilación iconográfica de vistas de todos los lugares que visitaba*" ["the iconographic compiling of scenes from all the places I visited"] (Romero 18), including an identification, description, and interpretation of the content, and then categorizing them into documentary photographs or aesthetic photographs, without suggesting a hierarchy but a distinction and variety of type, highlights the power of this medium for him. Cajal found in the process and products of photography a scientific technology at the service of humans on many levels, one that enhanced life, provided knowledge about it, and (sometimes, poetically) safeguarded it against the ravages of time. As science moved ahead in its quest for knowledge, and photography developed in accuracy, human beings and landscapes changed radically as well. He concluded that the latter is the most powerful impact of the innovation of photography: "privilegio de la fotografía, como del arte, es inmortalizar las fugitivas creaciones de la naturaleza. Gracias a aquélla, parecen revivir generaciones extinguidas, seres sin historia que no dejaron la menor huella de su existencia. Porque la vida pasa, pero la imagen queda" ["Like art, photography has the unique privilege of immortalizing the fugitive creations of nature. Thanks to photography, long-dead generations, those without history because they left not a trace of their existence, are brought back to life. This is because life ends, but images remain behind"] (cited in Romero 2). A slide may invite questions; an image evokes both affect and history.

It is telling that his grief over the death of his beloved son San-tiago, diagnosed with a fatal heart condition after earlier suffering from typhoid fever, coincided with the therapeutic exercise of the will that his book on color photography represents. He dedicated the volume to the memory of his son and revealed that "pensando en él inicié los primeros capítulos de este libro" ["with him in mind, I began the first chapters of this book"] (cited in Laín Entralgo and Albarracín 174). After noting what he referred to as this personal side of the project, Cajal returned quickly to the science of photography. He asked him-self about the potential value of the book for those who desired to study how to reproduce the colors of nature, done "*a través de tres procedimientos: indirecto, directo y, dentro de éste, interferencial*" ["by means of three methods: indirect, direct and, within the second, interferential"] (cited in Laín Entralgo and Albarracín 174). Within the short space of his remarks, Cajal twice used the words "valor" and "mérito" (174) as part of rhetorical questions addressed to himself about the project. Each time, the answer was yes, and he proceeded to explain why each step contributed to the learning of color photography by those interested in this new technique. In addition, he was careful to point out that what he set out to do he had accomplished, and more: "Aparte el cumplimiento de los objetivos que en su introducción me propuse, creo que es una exposición muy completa y fruto de detenidas lecturas, de los métodos heliocrómicos" ["Besides fulfilling the goals I set out in the introduction to be accomplished, I think this is a very complete presentation—the result of my meticulous and detailed readings—of the methods of color photography"] (cited in Laín Entralgo and Albar-racín 174). Scientific curiosity, thorough professional preparation, and delineated goals unite—in what the author called "*nuestra modesta con-tribución*" ["our (my) modest contribution"] (cited in Romero 2) to this emergent field—to outline, explain, and interpret practical means of photographing colors, "*trocando la siniestra visión de buho por la riente visión de hombre*" ["exchanging the dark and sinister vision of the owl for the happy, laughing vision of man"] (cited in Romero 2). What had been an unreachable ideal had progressed, through the inventions of the Lumière brothers and the limits of black and white, to within reach of the most modest photographer. For Cajal, in order for that amateur to do the best work possible he or she needed to be taught the steps and the shortcuts by a cutting-edge scientist such as himself.

Cajal's celebratory conclusions brought together the various pieces of the colorful kaleidoscope of his artistic and scientific work with an admonishing warning for those who may have considered photography a pastime, a hobby, or something to be picked up quickly rather than assiduously learned. He cautioned readers:

> la fotografía no es deporte vulgar, sino ejercicio científico y artístico de primer orden y una dichosa ampliación de nuestro sentido visual. Por ella vivimos más, porque miramos más y mejor. Gracias a ella, el registro fugitivo de nuestros recuerdos conviértese en copiosa biblioteca de imágenes, donde cada hoja representa una página de nuestra existencia [photography is not a commonplace sport, but rather a first-class scientific and artistic exercise and a felicitous expansion of our visual sense. Through photography we live more because we are able to see more and to see better. Thanks to photography, the fugitive record of our memories becomes an abundant library of images, wherein each leaf represents a page of our existence]. (cited in Romero 2)

The catalog, album, and cabinet now opened into the vaster space of a library in which Cajal and other photographers could record, collect, and compile all of the no longer fugitive images for future reference. Together, the images created and collected by *aficionados* through the turn of the century, and by physicians such as bacteriologist Jaume Ferrán i Clua and by cell biologist Cajal, in particular, joined scenes of sartorial styles, urban panoramas, modernizing capitals and their inhabitants with the art of penetrating *"en el misterio de la imagen impresionada"* ["the mysteries of the printed image"] (López Mondéjar 70). Perhaps the discoveries made under the lens of the microscope paralleled the emergence of the photographic image as activities that produce the visible from the invisible, and the totality from its fragments.

Just as the natural world was filled with wonders to sort out, so the emergence of an image from any one of the numerous photographic processes relied on the elements of chemistry that produced, in a sense, something from nothing. Kozloff asks, "what truth can be conveyed by records of sensitive faces when we know them to have originated only through the tiniest blip of light, abruptly isolated in past time?" (8). Benjamin might have added one cautionary additional note about the "tiny blip" and that is that it could become the spark—again deploying metaphors of light and chemical development—that for a single moment brings an uncanny notion back to the conscious level. As he wrote in one fragmentary text of the "Theses on the Philosophy of History," "The true picture of the past flits by. . . . The past can be seized only as an image which flashes up at the instant when it can be recognized and is never seen again. . . . To articulate the past historically does not mean to recognize it 'the way it really was' (Ranke). It means to seize hold of a memory as it flashes up at a moment of danger" (Benjamin, "Theses on the Philosophy of History" 255). The desire to retain that image, or the drive to piece together the historical whole, can point in the direction of myth and tradition (re-creation in situ) or establishing a connection with the current material world. The artistic photograph that captured that spark does not conflict with the scientific image that proposed to illuminate human knowledge. They coincide in not in recovering "what really was," but in shedding light on what may have been hidden or escaped detection by earlier observers.

A man of insistent curiosity and observation, of tradition and innovation simultaneously, a scientist interested in augmenting and expanding the visual with the aid of all possible technological apparatuses, Cajal's extant photographs can be divided into several categories. There are posed portraits of his wife, siblings, and children in accord with the conventions of the day. There are self-portraits alone and with family. He recorded panoramic scenes of urban life, and anthropological photos of the inhabitants of various regions of Spain and their daily activities. To this, he added an interest in monuments and architectural structures coming into being and structures being torn down. Of course, the laboratory where he spent his time on scientific inquiry appeared in many photographs. And he set up and photographed a number of still life compositions. He was overtly conscious of aesthetic or art photography and of the techniques that were being used to not just record people and places but to go beyond mechanical photography into the realm of light and contrast manipulation, the blurring of images for effect, and arranging or modifying the disposition of the persons and objects in pictures. It seems unthinkable for a man of principle in the area of scientific investigation to stray from that aim in other fields of endeavor, but it is evident that Cajal saw photography as a chance to simultaneously record detail, stage scenes, seek harmonious tonalities, pose individuals, introduce soft focus, and otherwise use the techniques of the photographic process all within the bounds of "the grammar of art" (Henry Peach Robinson cited in Kingsley 75) and not merely the grammar of science that contributed to the development of the image. The challenge of doing photography firmly bonded a hypothesis with an experiment, a theory with practice. For Cajal, this was an art just as the recording of observations was. On his work in color photography, he wrote

> El primer motivo fue contribuir, con mi modesta iniciativa, a divulgar entre los aficionados a la heliocromía los principios físicos fundamentales de esta maravillosa aplicación de la ciencia. Así lo expresaba en el prólogo que encabeza la obra. "Privarse de la teoría—decíamos—es desdeñar la mitad del placer fotocrómico, que consiste en comprobar experimentalmente la exactitud de los principios científicos. El devoto de la fotografía del color no debe ser rutinario practicón, atenido meramente a recetas y formularios, al modo del carpintero, que, aguijado por la necesidad, abandona la garlopa por el objetivo. Sólo acierta quien sabe. La interpretación de los resultados obtenidos y el remedio de los accidentes y fracasos, encuéntrase exclusivamente en la clara comprensión del mecanismo fisicoquímico de cada operación fotográfica."

> [The first motive of my modest undertaking was to contribute to the knowledge of devotees of color photography about the fundamental physical principles of this marvelous application of science. This I stated in the preface to the work: "To deprive oneself of theory . . . is to disdain half of the pleasure of color photography, which consists in experimentally testing the precision of scientific principles. The cultivator of color photography should not be a routine practitioner, merely adhering to instructions and formulas, like a carpenter who, moved by necessity, abandons the plane for the lens. Only those who know are successful. The interpretation of results obtained and the remedy for accidents and failures is only to be found in a clear understanding

of the physio-chemical mechanism of each photographic operation."] (*Recuerdos* II, ch. XXV)

Cajal's motivation to correct the nonsense of otherwise educated engineers, lawyers, and physicians—"cómo desbarraban" (*Recuerdos* II, ch. XXV) in public—spurred him to compose a volume of methods and procedures for even the more common reader, one illustrated step-by-step. Knowing how in photography the grammar of the art was as critical to him as the skills of the carpenter or, for that matter, the expert eye of the scientist. Having been trained how to see, how to propose a hypothesis and proceed from theory to empirical proof, and how to comprehend and remedy failures of the scientific process Cajal reflected the integrative drives toward clear evidence and toward the accurate working of the photographic mechanism. The greater the knowledge of the science of photography, the better the image produced. As a *docente* or teacher in charge of instructing pupils in anatomy, histology, and microscopy, he had a parallel job in the education of future photographers and in the establishment of a scientific persona whose interests and expertise could encompass both types of laboratory: the histologist's and the photographer's. He called the lens of the microscope "*la ventana del ocular*" ["the eyepiece as a window"] (cited in Laín Entralgo 134), a notion reflected in his aspiration of giving Spaniards a window on the art of composing and developing the photograph through his manuals. In turn, their participation in such technologies would provide them with an eye on the world. Holmgren's nomination of Cajal for the Nobel emphasized, as noted previously, the scientist's "breadth and inclusive vision." Such a comprehensive method of considering phenomena, technologies, and representation must include the acquisition of knowledge necessary for using the eyepiece, deploying the camera's mechanisms, and communicating with students.

Photography was situated at the critical juncture between scientific knowledge and the "calibration of the eye" needed for understanding, and that "inner logic" of rising capitalism that the incipient market for visual culture would present to the masses. If we compare several photographs taken within the walls of the home, it is evident that the artistic image of domesticity was reproduced in the details of Silveria who—in her work clothes and apron in one image—leans on a table with a tired look, gazing off-camera into the distance. At what does she look? What are her thoughts? What of the time involved in setting up the shot, waiting for its completion, and remaining still? In fig. 2.5—*Retrato de Silveria*, the patterned wallpaper frames her figure as do the focus and exposure that cast light on her face and clothing amid the shadows, wrapping her in gentle darkness as she seems to contemplate something one cannot quite fathom. In the home of the scientist, what occupied her time or what dreams did she postpone or displace in favor of the demands of the family? Was she merely concerned over the next chore to be accomplished, or did she share Cajal's valuing of this photographic record? She appears in work clothes—apron, buttoned shirt, and heavy skirt—with her sleeves rolled up. Her hair is less carefully pinned up than in other images, indicating not leisure but active engagement with something. This photo was taken during a hiatus, a break in time when Cajal left the laboratory for the dark room and she left her routine for his art.

In contrast to the domestic space within which the woman works, an entirely different take on the home shows the scientist whose vision produces innovative modern laboratory and darkroom results for Spanish culture. Seated at a table laid with a colorful cloth, Cajal may also gaze off into the distance, but it is a gaze focused on the human skeleton poised alongside and not the downcast glance of his wife. He has taken a moment of time to be recorded as he studied for examinations, preparing himself for a bright future in science. His preoccupations are not, and cannot be, hers. Both husband and wife rest their heads on their hands, one looking upward toward the intellectual ether and one downward to material concerns. Silveria has two books of some sort piled near her elbow, on a table covered with a cloth; Cajal has thousands of books that constitute a library flanking him. He posed before shelves and equipment; she sat down to rest. The books next to her are props; his are indicative of learning, promise, and accomplishment. As Cajal used the family table for his scientific work, Silveria paused for an instant at a table for her husband's photograph. The same table acquires different meaning when represented in the presence of two distinct types of people. These are two worlds juxtaposed: a domesticity reminiscent of traditional values, and the intervention and interruption of scientific thought into the space of the home by means of both objects and imaginaries.

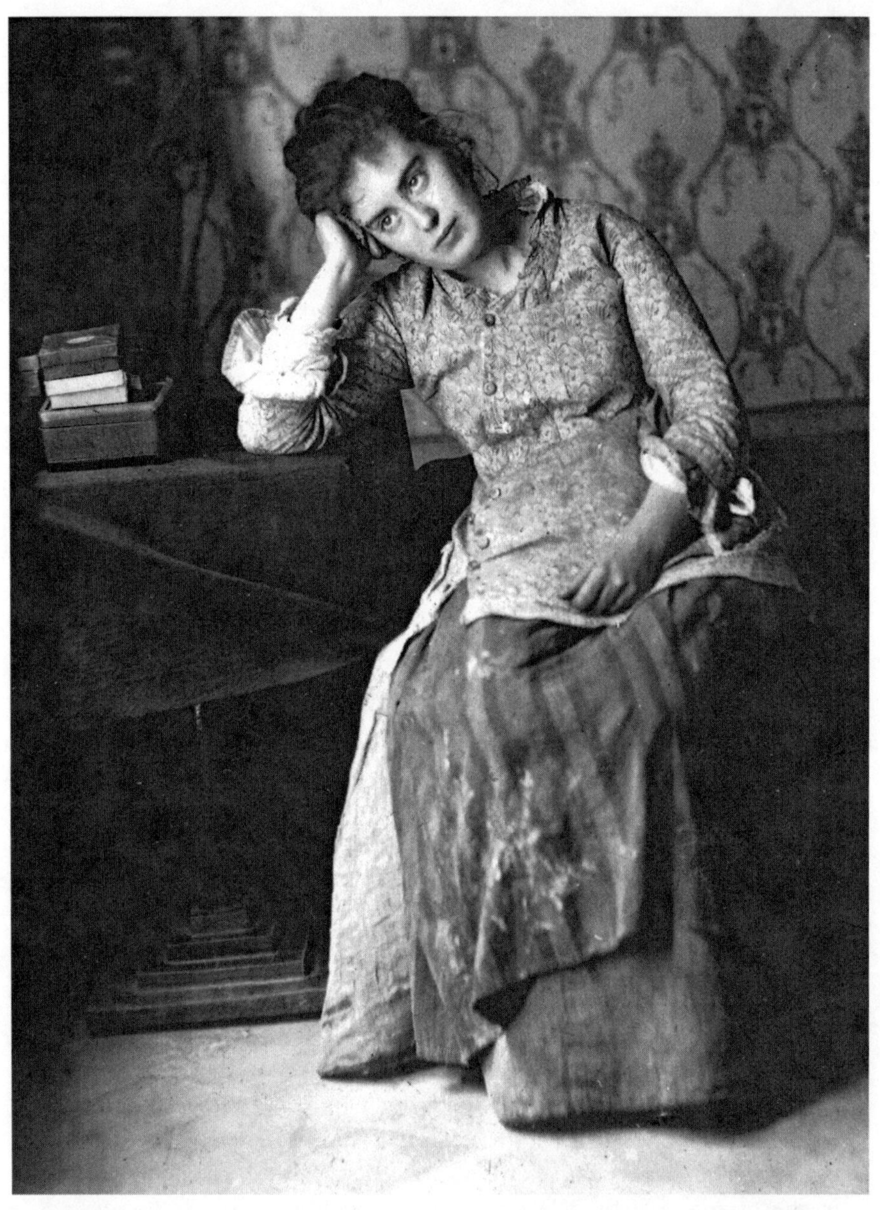

Figure 2.5. *Retrato de Silveria* [*Portrait of Silveria*]. Santiago Ramón y Cajal. Legado Cajal (CSIC). Instituto Cajal. Madrid, Spain. Used by permission.

Like Charles Darwin some decades before, Cajal seemed very aware as well of two communities of readers to which he addressed his written texts. For the first, their peers in the scientific world across Europe, the men of laboratories and investigations could adopt a tone conforming to the basic standards of objectivity, the value of observation, and the presentation of hypotheses and evidence. Argumentation based on reason accompanied a study of particular phenomena of the natural world to create what James A. Secord finds as one of Darwin's central points in his *Recollections* (published in 1887 after his death): "his lifelong attempt to create a new scientific persona" (*Evolutionary Writings* xxix). Similarly, Cajal in *Recollections of My Life* has a two-pronged impetus: to bring into focus his own investigations of the natural world, and to bring to the forefront of research around the world his name as a representative of scientific work being done by Spaniards. For the second group of readers, a language more suited to a general readership is required in order to provide what today is called a popularizing of scientific knowledge, research, and innovation. Far from defending a superficial notion of science—whether exploring cell biology or color photography—Cajal's mission was to educate, to "meet cultured people and to be able to practice photography and chess, hobbies, he said, 'where you do not bet money but your brain, our greatest capital asset'" (Ramón y Cajal Junquera 59). It is telling that he did not refer to the market for photographs but to the processes of photography as intellectual assets (capital). Given his intellectual perception of even a "hobby," Cajal elevated chess and photography to a higher level. All of these activities stimulated the brain, challenged the intellect, and provided information about human beings and how their brains function. His interest in the physiology of the nervous system was complemented by a fascination with psychology, both derived from his discoveries about neurons and how they worked to link the inner world of the human with the outside objective world. Cajal's comments on the "caminos cerebrales" ["cerebral paths"] (Ramón y Cajal, "Prólogo: Madrid, 22 de abril 1904" 80) that lead to faithful or entirely aberrant representations of the information gleaned about external things through the neuronal paths opened the door to his studies of logical associations or "*conexiones antinaturales*" ["totally unnatural connections"] (Ramón y Cajal, "Prólogo: Madrid, 22 de abril 1904" 81), both of which were worthy of analysis.

Using a metaphor of light and heat from domestic space, Cajal called these complex processes of association "*la combustión en el horno del pensamiento para la forja de relaciones causales nuevas, de conceptos superiores, de síntesis luminosas, de excelsas creaciones de la razón científica o de la fantasía poética*" ["the combustion in the oven of thought in order to forge new causal relations, superior concepts, luminous syntheses, sublime and lofty creations of scientific reason or of poetic fantasy"] (Ramón y Cajal, "Prólogo: Madrid, 22 de abril 1904" 81). As a whole, such dynamic psychic activity—memory, imagination, association, conscience, logic, and sentiment—deserves to be brought out from the "sombra" (Ramón y Cajal, "Prólogo: Madrid, 22 de abril 1904" 85) into the realm of scientific exploration and debate so that human beings might be the masters of their potential.

Those such as Cajal whose education—in his case, mostly self-education—in the chemical processes of photography could extend that field further into the cultural narrative of the time through a promotion of producing better quality images to the Spanish middle-classes, contributed to its broader popular dissemination. Darwin's narratives based on the notebooks from his voyage on *The Beagle*, as well as his *Recollections*, are reminiscent of Cajal's *Recollections of My Life*, writings on how to get the best results from the new color photographic processes, *Charlas de café* [*Thoughts from the Café Scene*], and his *Reglas y consejos sobre investigación científica* [*Rules and Advice about Scientific Research*], a manual of advice for young scientists that ranges from general methodological proceedings to how to strengthen the will, and from avoiding a paralyzing and excessive admiration for the great works of scientific innovators to the false dichotomy between theory and practice. In his photographic essays, Cajal was clear in his focus more on the serious apprentice and less on the professional. His discussions were constructed in the first-person singular and plural—I and we—in order to bring readers into the fold; his sentences frequently end in exclamations—four of them in a row when he reveals the "*gérmenes de grandes invenciones . . . en las obras de los antiguos*" ["seeds of great discoveries . . . in the works of the ancients"] (Cajal, *Reglas y consejos* 40–41), especially referring to the philosopher Seneca from whose work on the magnifying power of a glass filled with water he finds early evidence of the same inquisitive spirit as the scientist had with the modern microscope and the telescope, biology, and astronomy (41). The most convincing evidence of Cajal's directing his comments toward those just entering a professional world comes at the end of a section on the false concept of a lack of new scientific themes to explore. He summarizes, to encourage any and all questions as worth pursuing: "En resumen, no hay cuestiones pequeñas; las que lo parecen son cuestiones grandes no comprendidas" ["In conclusion, there are no small matters; those that appear small are in reality large matters that are not yet understood"] (41). The small issues—what had always been the invisible details unexplored or unexpected, made accessible to the human eye through the microscopic lens—offer another window on the natural world through which one might begin the investigation of those "larger matters" whose enigmas would be more difficult to answer, even with all the technological resources at society's disposal. While a resolution of discrepancies and discontinuities may have appeared at first to be the goal of making society and nature visible, this directly conflicts with what Benjamin explored as a clash of codes of experience with the advent of modernity, a clash that generates anxiety in both the producers of images and their consumers. Trying to make sense of "history [that] is less a narrative than a series of visual moments or scenes which inescapably permeate the present as well as the past" (Larson and Woods 1), the convolutes of space, time, and the absolute complicated more than they straightened things out. Despite the lure of the camera and the microscope, those apparatuses do not signal an end to inquiry but the opening of inquiring activities.

Carlos Monsiváis writes of the rise of photography in Mexico in terms that can be applied to what López Mondéjar calls the "*democratización del retrato*" ["democratization of the portrait"] (51) also beginning in the mid-nineteenth-century in Spain, with social changes linked to economic ones, all tied together by technologies of travel and social mobility. Monsiváis summarizes: "La fotografía: negocio, innovación técnica, curiosidad adulatoria, nicho de la devoción por los rasgos amados, desafío a los olvidos y a las inclemencias del tiempo" ["Photography: business, technical innovation, flattering curiosity, niche for the devotion to beloved traits, defiance of forgetting and of the inclemencies of time"] (15). When photographs are simultaneously records of the faces of loved ones, commercial sales objects, displays and exhibits, and antidotes against the passage of time, they encapsulate the complex experiential moment of modernity. The business aspect of photography can be discounted in Cajal's case, since he did not cultivate the art for purposes of income but for the recording of the relationship between a father, a man of science, and his family. The survival of these images taken multiple times did suggest an answer the question of his sentiments toward loss and the passing of time, as well as toward the desire to document and perpetuate the face of a social class from within. The simultaneous acquisition of camera equipment and making of images reflects a confirmation of their value. In his private collection of over five hundred extant photos (what is left of the thousands he is said to have taken most of which have been lost or dispersed), the images respond to his theories on photography previous to, and later published in, the magazine *La Fotografía* beginning in 1901. The description of photography as a serious process that had to be learned, as a series of challenges to be overcome, as an art to be cultivated over time. All of these aspects contributed to recording, remembering, documenting, and adding to a science that was able to mimic and augment what the human eye could perceive for the continued intellectual and social life of the community. These photographs responded to conventions operative during the second half of the nineteenth century and the early twentieth, first and foremost in the realm of the portrait and self-portrait.

Publio López Mondéjar recapitulates:

> Junto a la fotografía de viajes y la reproducción de obras de arte, el retrato se convirtió en la manifestación emblemática del desarrollo imparable de la fotografía. La apoteosis del retrato se produjo en una época en la que, paralelamente, se estaba operando una profunda mutación social. El decidido impulso de la burguesía liberal decimonónica fue postergando a las viejas castas políticas y aristocráticas, y la consiguiente democratización acabó propiciando unas formas artísticas más accesibles a las nuevas clases sociales en ascenso. Hacerse retratar se convirtió en un signo de progresión social.

> ["Together with travel photographs and the reproduction of artworks, the portrait became the emblematic manifestation of the unstoppable development of photography. The

consecration and glorification of portraiture arose at a time during which a profound social change was simultaneously occurring. The determined impulse of the nineteenth-century liberal bourgeoisie was replacing the old political and aristocratic castes, and the subsequent democratization of society ended up favoring artistic forms that were more accessible to the newly ascendant social classes. To get one's portrait taken became a sign of social progress."] (51)

The camera as *the* device that symbolized modernity was able to record the look of "anxiety" (López Mondéjar 60) of those living the modernizing moment, but it also made visible their reliance on the photograph as material evidence of belonging to that time of "progress." Technological modernization and cultural modernity came together in the device and in the look and composition of the photo it produced. Albums were filled with copy after copy of faces and figures to create a fascinating relationship between the fleeting life of human beings and the burning desire to leave a record of clothing and gestures, siblings and parents, homes and places of work, scientific devices and discoveries. The sharper the image of those faces and figures, the more they revealed the pace of the times, the greater the concurrent triumphs of science and art.

The chemistry behind the production of the image—with its "accidents" and fortuitous discoveries as Cajal himself documented, as well as the inventions of others of which he learned—and the use of multiple-lens cameras, or the perfecting of several exposures on one photographic plate, invited the proliferation of the *carte de visite* or postcard-like portraits similar to what Cajal exhibited in his collection as multiples of the images of himself and his family (see fig. 2.6—*Familia* [*Family*]). In these two frames, the four children exhibit two fundamental differences. First, the lighting between the two shots changes dramatically from the first to the second photograph. It is not certain but perhaps Cajal was experimenting with this element after seeing the first, darker product. The second detail is the expressions on the children's faces that register changes, as well as the nuances of their pose. In the first instance, three faces are evident while one child is held in her sister's arms and less visible to the camera. The second round produced smiles, all four seated—perched, more or less—on the

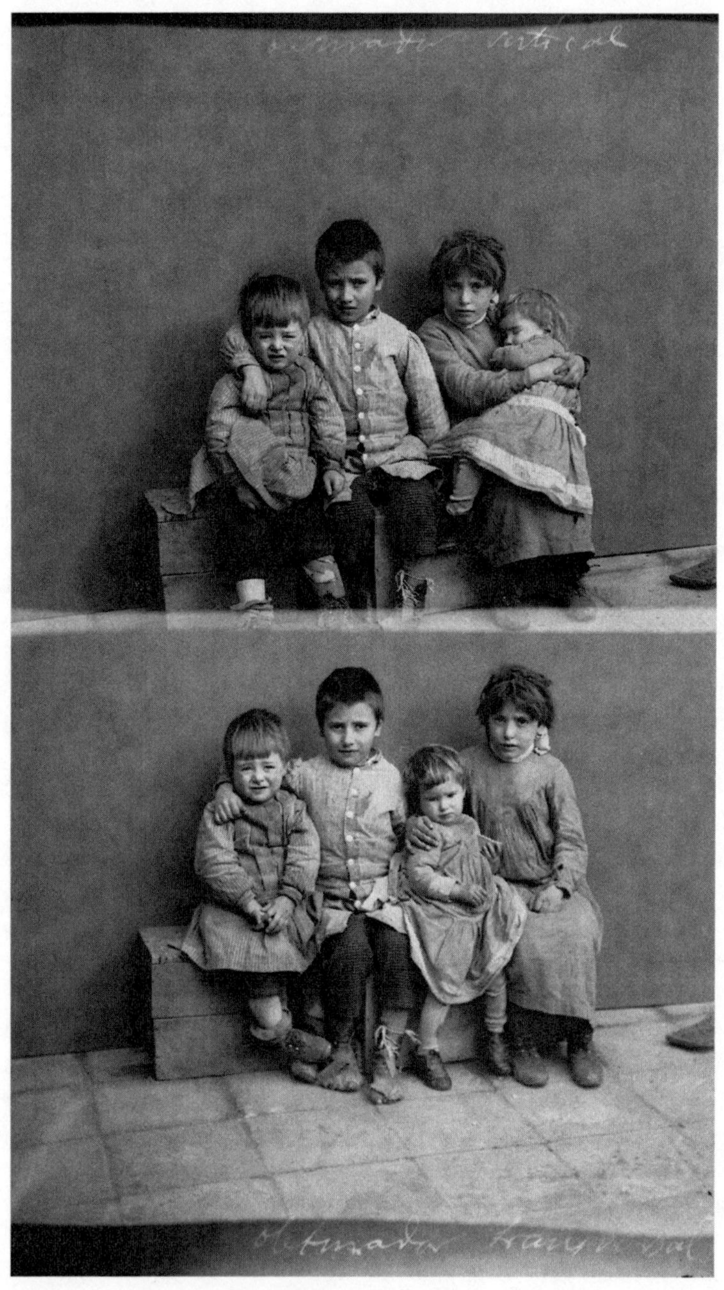

Figure 2.6. *Familia* [*Family*]. Santiago Ramón y Cajal. Legado Cajal (CSIC). Instituto Cajal. Madrid, Spain. Used by permission.

bench and maybe the indicators of restlessness after having sat through one photograph already.

Cajal's multiple and repeated family photographs reveal the grammar of composing an image: the gestures allowed, the look expected on the face of the adult or child, the presence or absence of emotion, a frozen image or the capture of unexpected movement. Kozloff writes that the "the stiffened bearing of personages reflects not only the long exposure time required for the photograph but also the decorum required . . . the sitters are viewed as dramatis personae engaged in representing a story about themselves, or else they are glimpsed in the process of a story" (26). How long one might pose, or how many times an image can be made from a single sitting, were indexes of the success of the photographer as well as of the personal qualities of the person portrayed. Images of his children were like portraits of the photographer's own reflected childhood. They comprise an album of "*la memoria de la especie*" ["the memory of the species"] (Monsiváis 23), an accumulation of familial and national values and not merely the faces of singular individuals. One such value reproduced insistently by Cajal was the pose of the child as the portent of the adult. How a child would be incorporated into society as a mature individual might be insinuated from the portraits of his or her youth. Respectable adults had been respectable children and respectable women had portraits taken as wives and mothers. The fact that his seven children sat for multiple photographs—one surmises that he insisted on it—brought to light family patriarchy as much as the potent rule of the scientist. Cajal provided us with a series of family portraits confirming the faith of the photographer in the outcome of his progeny through the evidence presented in the photographic document. He cannot predict the untimely death of his son Santiago as they posed for the earliest shots, but that event will come to disturb the chronology of success and unity predicated at the outset with all the children present. The family will indeed go on, but the failure of science to prevent death, to cure the results of disease, was a pall over the final scenes much as the fading of his sight in old age was a fear that would keep Cajal preoccupied.

Cajal's collection of photographic portraiture work was no exception to the conventional standards, although they showed little of the regalia of more formal poses of the time. Aside from the studio portraits of the scientist in his academic robes, most photographs contained a similar likeness: the professional man in a suit. A black-and-white portrait of Cajal and his wife, when placed alongside a later color photo of the couple (fig. 2.7—*Retrato matrimonial* and fig. 2.8—also labeled *Retrato matrimonial*), reflects the traditional poses of husband and wife, with the constancy of their formal relationship documented through time. His attire reflected the professional man, with or without a family of posed and stiffened individuals recognizable as exemplars of the values and codes of the day. The modernity at play in these images is not modern childhood, or a cordial relationship between spouses, but a father figure and a demure mother figure, alone or surrounded by their progeny. In the first instance, Silveria stands next to her husband and both are attired in formal clothing. She wears the dark dress seen in the earlier formal family photograph taken by Cajal, and holds a fan in her hands. Her wedding ring is visible, but that just confirms their relationship that is evident from the portrait. Her hair is well groomed, not unkempt as in the photo taken with Jorge and Pilar, perhaps posed in the midst of everyday turmoil, and the couple has a look of propriety and decorum.

The second portrait of the couple differs in its change in the conventions of such photographs as well as in the tones of color made visible on Silveria's more modern dress. The reserve of the husband and wife facing the camera in fig. 2.9 has been broken and they now turn a bit more toward one another. Cajal is still seated, but he now holds his wife's forearm as she places it across his shoulder. The man of science's suit is still formal but the vest is much lighter than the previous image shows, and the pants have a faint stripe. They would be more appropriate for the city than the country since urban life, despite any dirt in the streets, would be less prone to take its toll on such vestments. It is Silveria's wardrobe that shows the greatest change, however. First and foremost is the light pattern across the teal-colored surface of her dress, but there is now lace draped over the shoulders and down the sleeves to her wrist. The collar of the dress is less stiff and high, and her throat is a bit more exposed. There is little doubt that he is the center of the focus as well as the center of the relationship, since he may be gazing into the distance but she still looks at him. The oval frame of the image enclosed them in a domestic scene of calm and well-being, now enhanced by color and tone that add richness to their image.

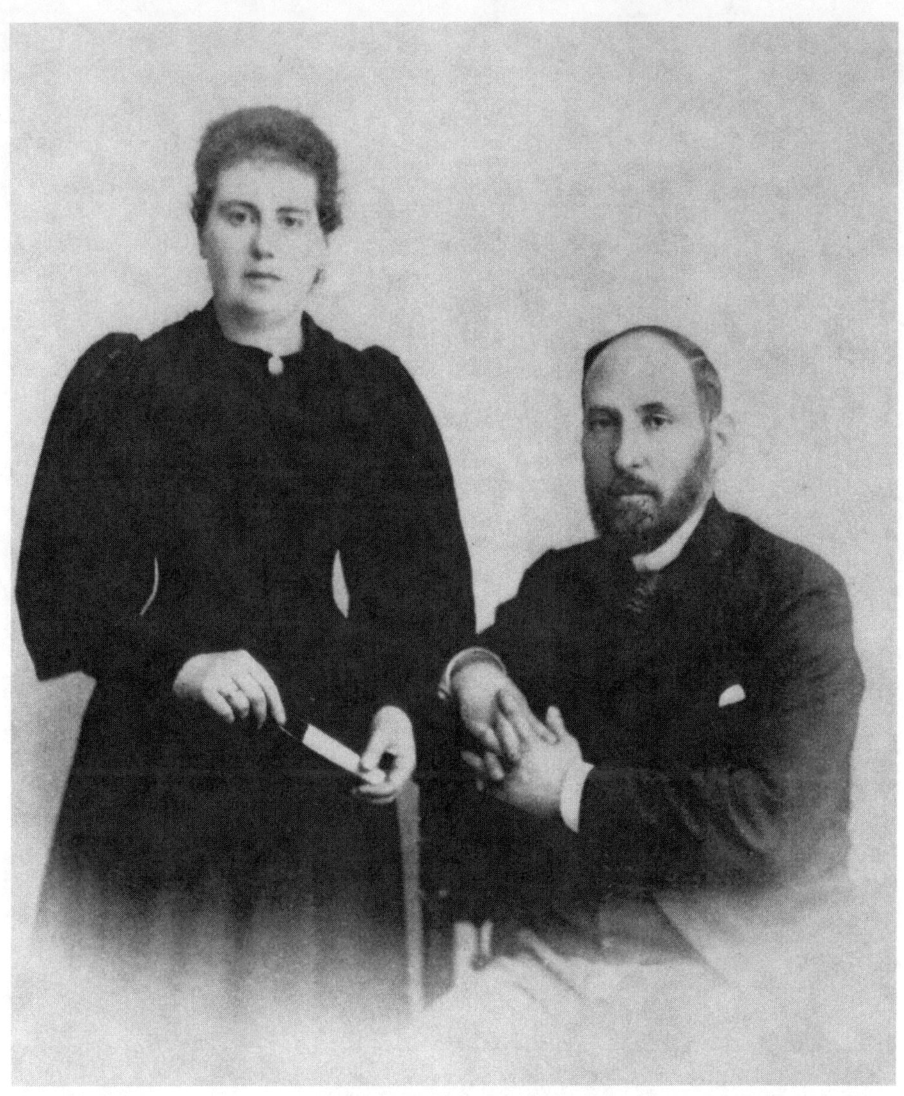

Figure 2.7. *Retrato matrimonial* [*Wedding Portrait*]. Santiago Ramón y Cajal. Legado Cajal (CSIC). Instituto Cajal. Madrid, Spain. Used by permission.

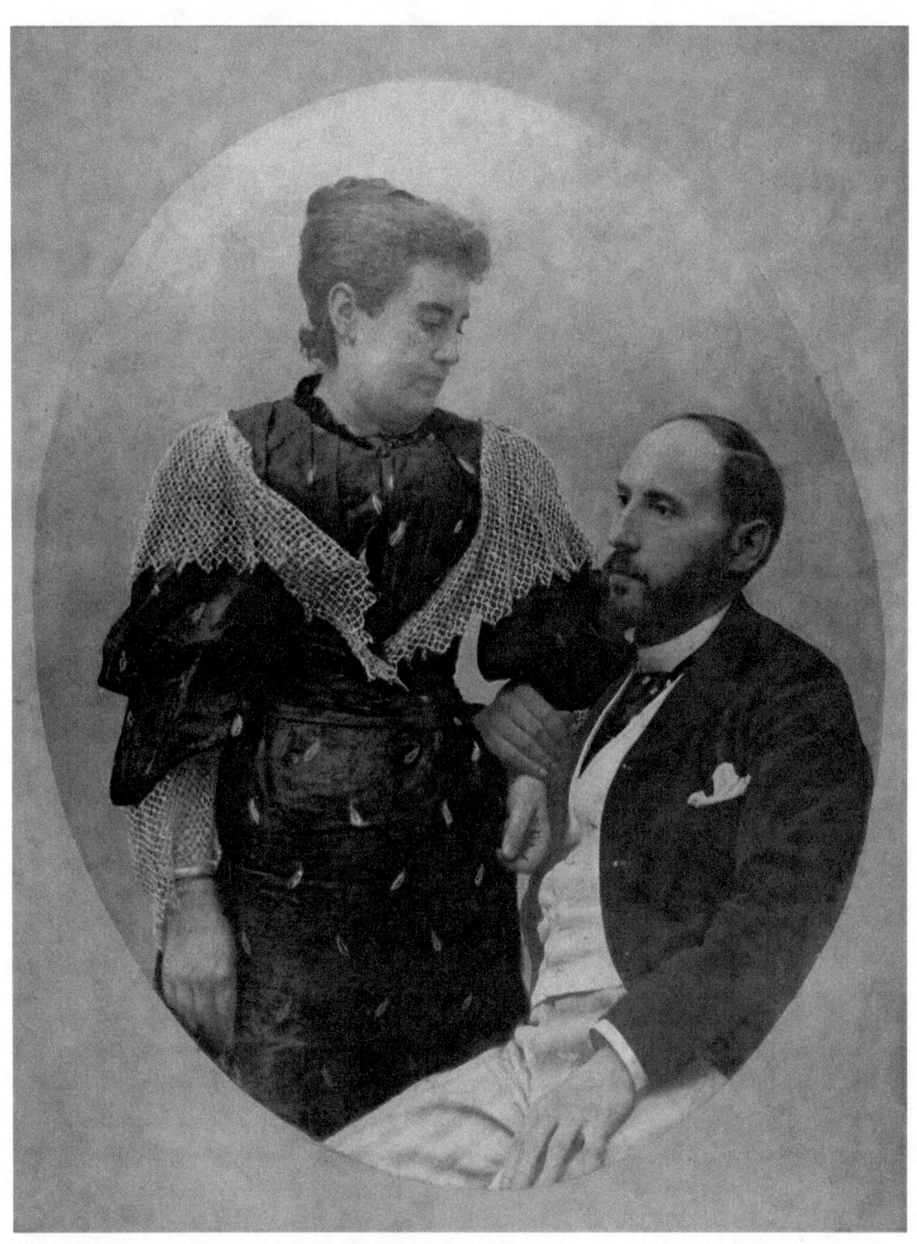

Figure 2.8. *Retrato matrimonial* [*Wedding Portrait*]. Santiago Ramón y Cajal. Legado Cajal (CSIC). Instituto Cajal. Madrid, Spain. Used by permission.

Looking at fig. 2.9—*Retrato: Silveria, Pilar, Jorge*, an early photo of wife doña Silveria shows her with Jorge and Pilar, their first two children, held tightly in her arms establishes the social roles of mother and child, and the power of the lens to convey and commemorate this relationship. One could even imagine this portrait subtitled "Motherhood" as Cajal has blurred and faded the background (the muted, angular shapes seem to be the walls and doorway of an interior room, but that is conjecture based on other photos), so that hazy shadows of dark and light fall gently behind the closer focus on the mother and children in the foreground. What matters most is most visible, not blurred. The setting may have less impact than the figures facing us because we cannot decipher it entirely, and our eyes are drawn to people over place. Silveria is seated on a simple, solid chair, her eyes closed in dreamy motherhood, wrapping her arms gently around the two babies in a light embrace. The strength of her arms signals she will not let them fall or move no matter how precariously they are posed. The children are dressed in clothing typical for their ages, from knitted headwear to over-the-ankle boots, and only Jorge gives any indication of a smile. He holds his fingers over his mouth, perhaps reluctant to remove them from his face as he has just been told to do. The photograph is not filled with emotion, but something of a repressed giggle is hinted at although not truly defined. They are snapped in limbo, between the scurrying to sit in their assigned places and their desire to slip away again. Like time, there are moments preceding this scene and moments that follow, all of which add up to what Barthes calls "the very tension of History, its division. History is hysterical: it is constituted only if we consider it, only if we look at it—and in order to look at it, we must be excluded from it" (65). Cajal is part of the shot in his role as photographer, as well as the observer of it, a contradictory and tense situation. Together, the three figures in the image form a portrait of a family in happy contemplation of reproducing the values associated with a rising social class. After all, they have both the financial means to be in good spirits and the scientific knowledge to produce such portraits. The photographer has convinced them to sit still for posterity and for his own testing of the camera.

A similar portrait of mother and children—this time, the children are Paula and Jorge—has Silveria looking down as she sits holding her daughter, while Jorge smiles from a tiny chair at her side. Silveria's

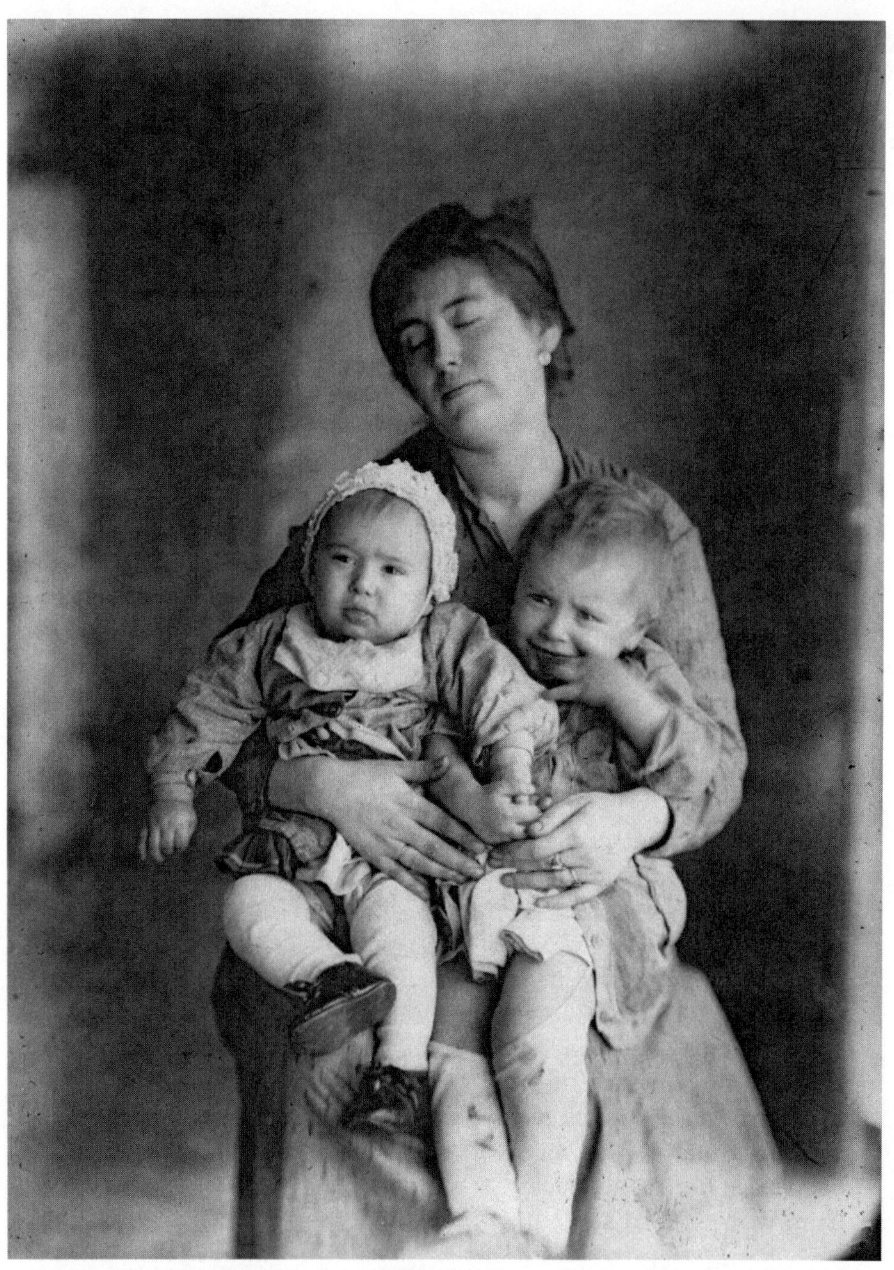

Figure 2.9. *Retrato: Silveria, Pilar, Jorge* [*Portrait: Silveria, Pilar, Jorge*]. Santiago Ramón y Cajal. Legado Cajal (CSIC). Instituto Cajal. Madrid, Spain. Used by permission.

downcast eyes tell us that she should not look at the camera, whether by her own volition or by direction, but that she is aware of being photographed as part of a family portrait. She has not been caught by accident; she is not frozen in the middle of movement, but posed demurely and decorously. They have not been surprised in informal attire, but in standard clothing for this social class; nothing less will do for the image as it records futurity and success, not failure. This time, the backdrop is not blurred, and walls, balustrades, and a stairway are visible to make a comfortable domestic scene complete and familiar. Numerous other photographs of the children, sometimes without their mother and a few times with their father who holds the long trip wire that allows him to snap the shot with himself included, are equally composed and quiet (see fig. 2.10—*Familia* [*Family*], fig. 2.11 (children alone), or fig. 2.12 (Silveria and the children). The objects that surround them and indicate their location vary, but all are domestic scenes. Sometimes a doorway appears as an unfocused mass of dark wood opening onto another room; other times, wooden chairs indicate a living or dining room has been rearranged to facilitate the shot. All indicate spaces of home life and the value of family, repeating the scene multiple times as if to preserve it untouched, but, in the process, revealing a need to reiterate gestures, stances, and mannerisms to convince an observer of their permanence and languish in it. Instead of this, what is established is transience—the early presence, then absence of Santiago, for instance—and contingency. In other words, figures embedded in moving and changing history are evoked as anecdotes. A complete narrative story is impossible despite the hand of the photographer being firmly in control of the remote shutter.

A series of photos that included four children—portraits taken with his German Steinheil camera with two exposures per plate—captured both permanence and change (time). In the first frame of one, the four are posed on a low bench against a plain wall, with large ceramic tiles covering the floor. Paula holds Fe as if she were practicing for motherhood, and the boys put their arms around one another in fraternal comradeship. There is sibling camaraderie, but not a lot of emotion. The second exposure of the shot shows Paula and Fe side-by-side this time, and the two boys interlace their feet as if they are going to jump up as soon as the photo was taken. Another set of two side-by-side photos in fig. 2.11 includes Jorge, Fe, and Santiago inside

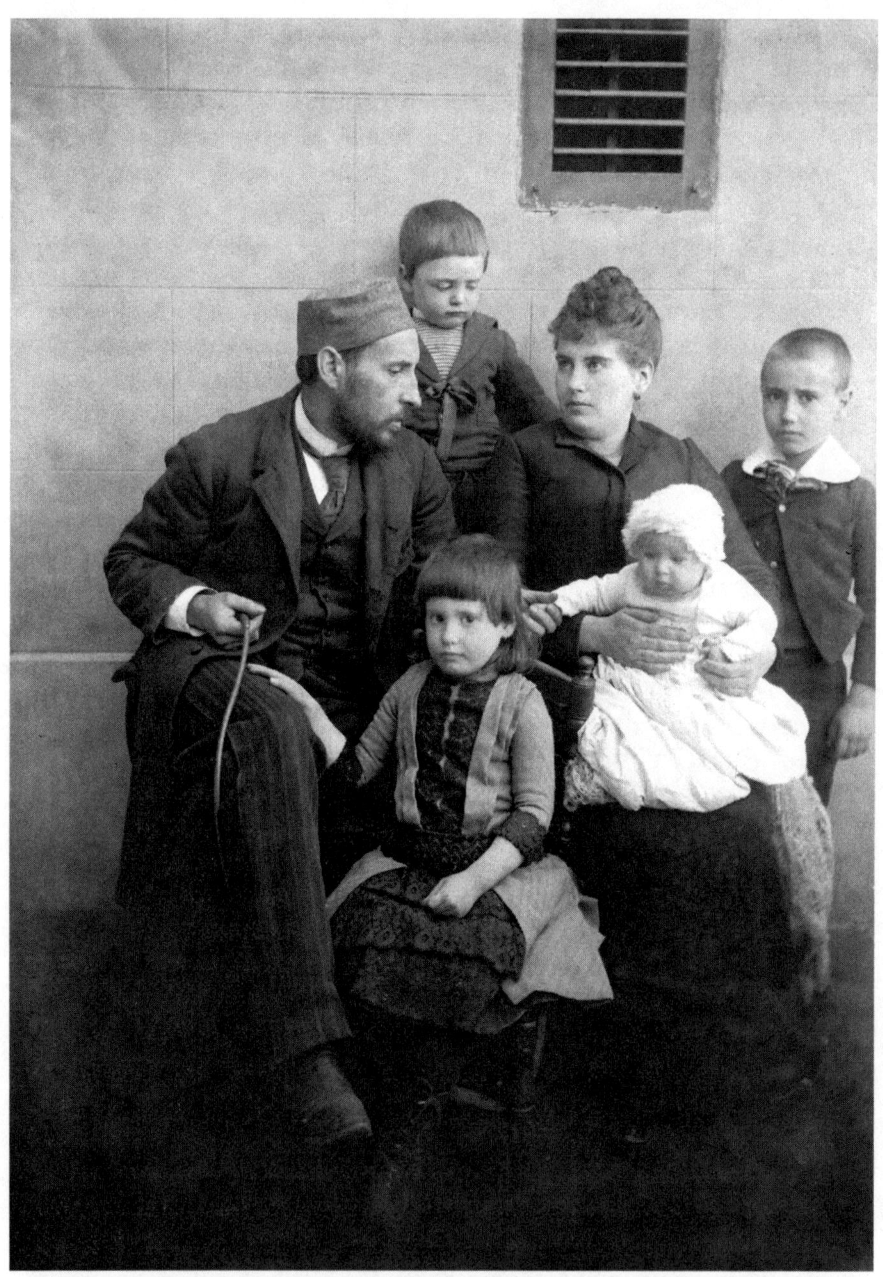

Figure 2.10. *Familia* [*Family*]. Santiago Ramón y Cajal. Legado Cajal (CSIC). Instituto Cajal. Madrid, Spain. Used by permission.

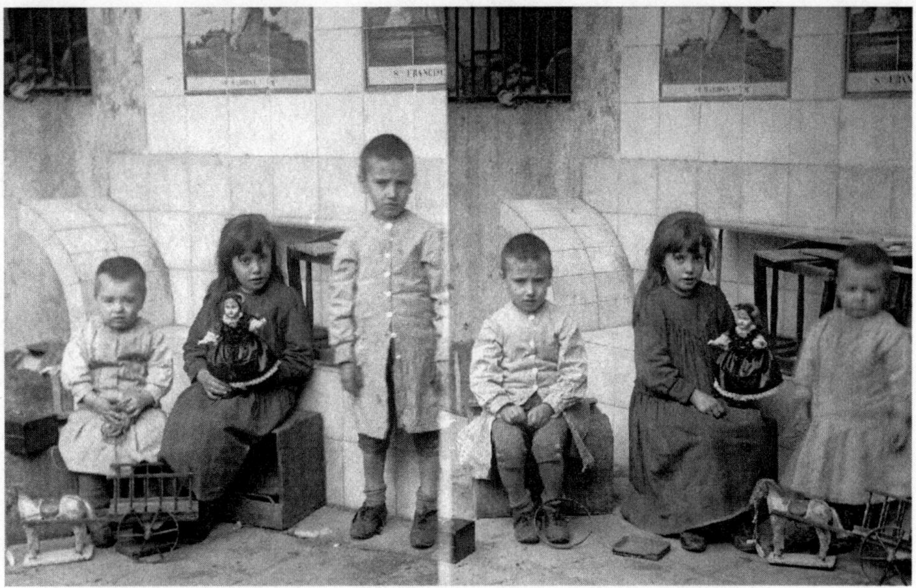

Figure 2.11. *Familia (retrato doble)* [*Family (double portrait)*]. Santiago Ramón y Cajal. Legado Cajal (CSIC). Instituto Cajal. Madrid, Spain. Used by permission.

a partially tiled room alongside a wooden horse and a doll. Just within view is a set of images of saints on the wall above them, and a chair and small table behind them. It is the only time they look like modern children, with the props that accompany their play surrounding them and attesting to their activities off-camera. Finances are good enough to purchase such toys and entertain the family appropriately; they do not work neither do they perform chores. The four appear protected within the walls of the house, and all stand still for their father's snapshot. And they stand still twice. The blurring on the right hand side indicates movement, whether the normal restlessness of a young boy or the desire of someone who has posed for enough photos already to return to carefree games.

Another portrait of the family in fig. 2.10—four children, mother, and father—becomes a simultaneous self-portrait as Cajal placed himself in the picture using a wire mechanism to trip the shutter at a distance. The close proximity of parents and children, the wife looking at her husband instead of at the camera, and holding the baby Luis

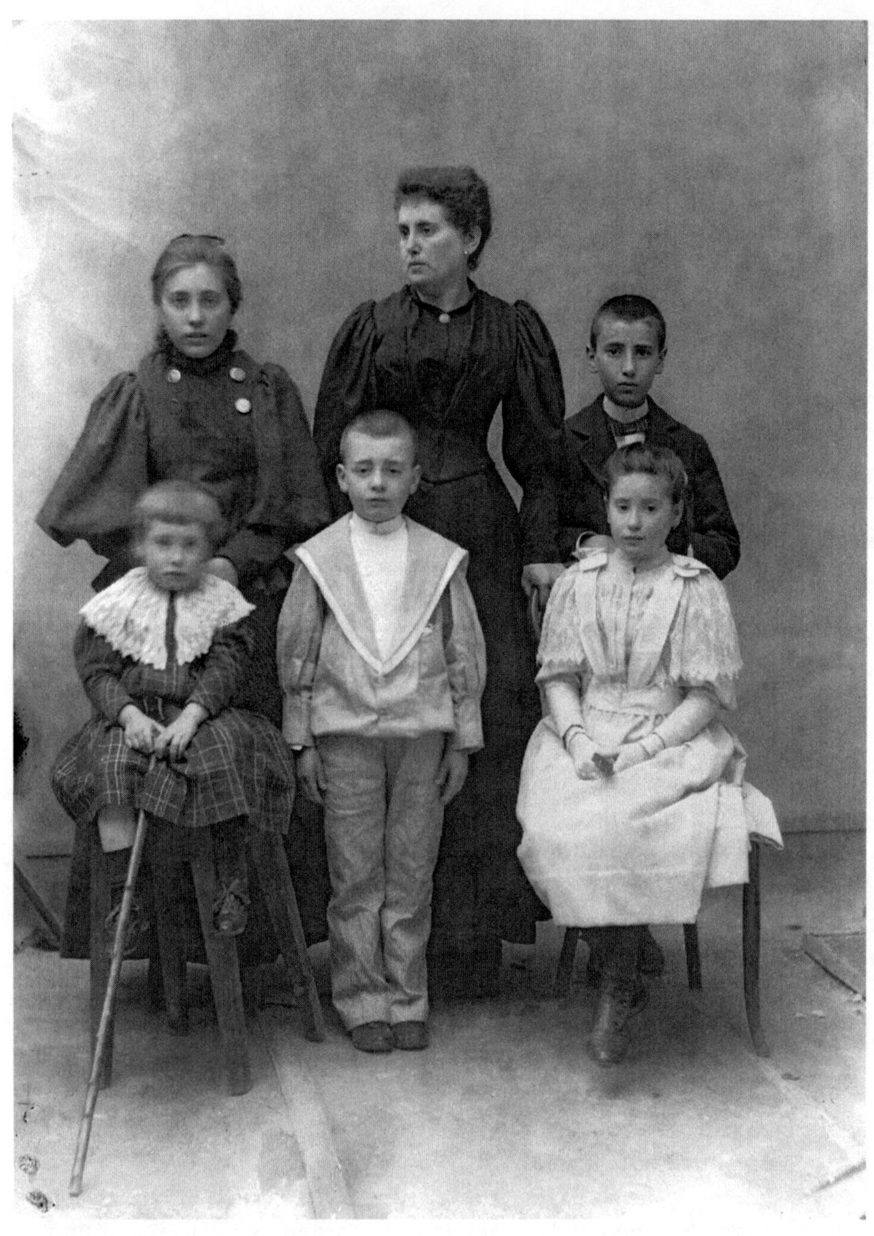

Figure 2.12. *Familia* [*Family*]. Santiago Ramón y Cajal. Legado Cajal (CSIC). Instituto Cajal. Madrid, Spain. Used by permission.

firmly in her lap, Paula's tiny hand resting on her father's leg, and Cajal grasping Luis's hand give the photograph a bit more of an emotional charge. Here the man of science is a man of artistry as a photographer, and a family man in the middle of things besides. Even as he presses the button to take the photograph, Cajal looks over to the others to check where they are and how they look. He was careful to compose what others would see. Not merely a face among others, he was also the person in charge of deploying the technology that recorded the scene and of staging its look.

While this scene confirms the materiality of the moment among relatives, his self-portrait of the 1870s, taken on return from Cuba, emaciated and suffering from a lung disease shows a pale Cajal framed in semilight, dressed in a three-piece suit but with a bulging handkerchief in his breast pocket, is pure contingency. That white handkerchief acts as a synecdoche for the coughing of the lungs to expel waste, a visible sign of an invisible disease. The white square also indicates time, for it relates to the whole of the stay in the sanatorium in recuperation outside the city. Science is connected here to the look of the man just as much as it is in the self-portraits sitting next to multiple microscopes, a microtome, poised with his hand on laboratory equipment, or with his students next to a cadaver. A simple piece of cloth protruding from his pocket links the man to a whole class of ill human beings, an entire group of voluntary doctors, and to the requisite treatment of this illness by scientific means. Through Cajal's use of the camera, there is an insistence on portraying the man of science as a patient of science.

Recognizable to the observer from so many portraits, doña Silveria is an interesting paradox seen through the lens of the modern camera: she is materially present but not the center of focus in most photographs. As one element in many family portraits, whether in the front row surrounded by children or in the back row standing behind them, she rarely looks directly at the lens. In an 1893 shot (fig. 2.13), five of the children surround Silveria, all dressed formally and either standing stiffly at attention—like Jorge in the front row and Santiago next to his mother—or seated calmly and quietly. Young Luis—the only slightly blurred image in the scene—holds some sort of walking stick, his sister Paulita has her hands folded in her lap, and Fe holds the back of the chair on which Luis is sitting or holds him still by

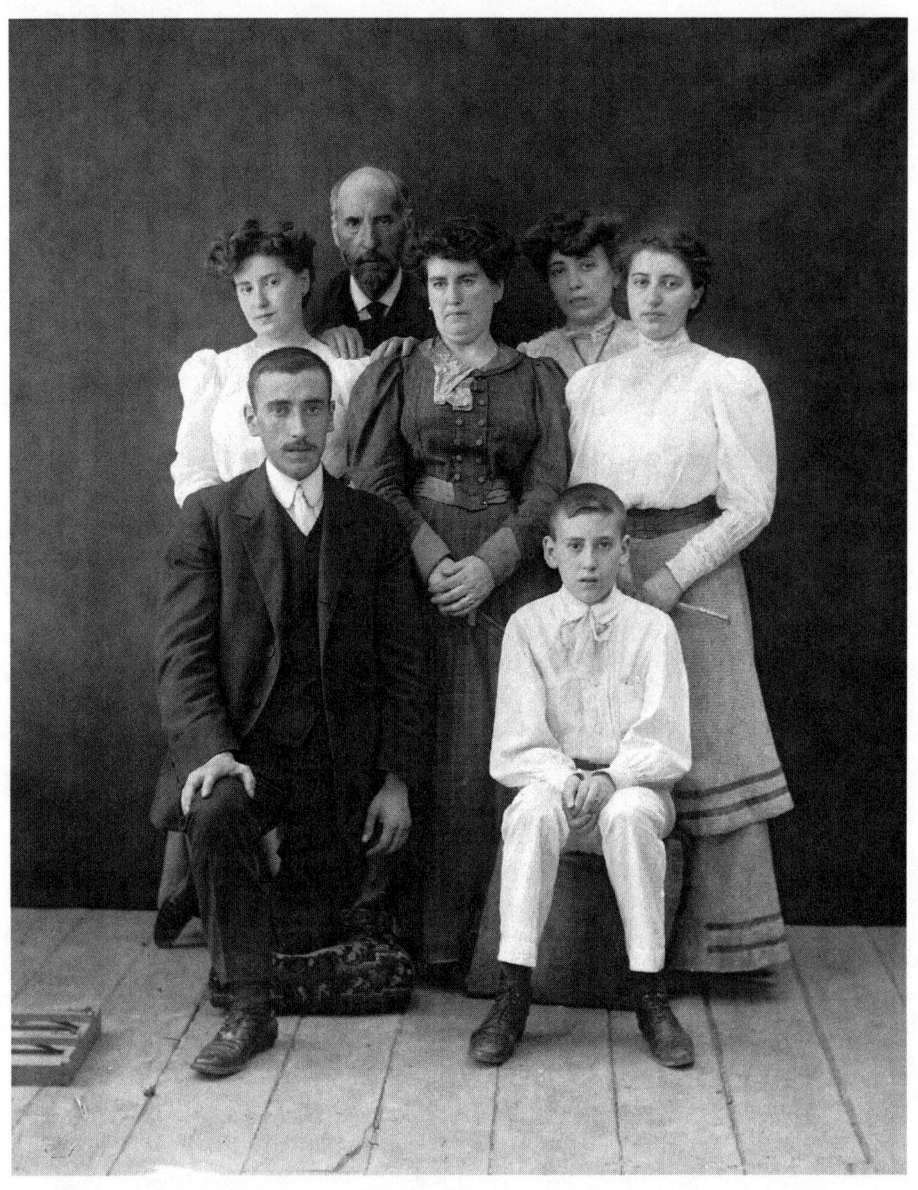

Figure 2.13. *Familia* [*Family*]. Santiago Ramón y Cajal. Legado Cajal (CSIC). Instituto Cajal. Madrid, Spain. Used by permission.

the shoulders for the photo. Five pairs of eyes stare straight ahead at the camera—and their father—while Silveria looks off to her right. An elegant woman, dressed in the long, dark attire typical of the day with a small brooch at the throat of a high-necked dress, she is the only one who does not make eye contact. Cajal's self-portraits may have him gazing off to one side, but that indicates thought and scientific introspection, since he is always shown accompanied by his many modern instruments of investigation. On the other hand, when photographed with her family, Silveria's role is to support her children and spouse, not to be the center of public attention and not to be meeting anyone's gaze.

In two portraits taken by Cajal, she does face the camera, but that admixture of innocence and melancholic resignation that he says first attracted him to her dominates the image. In his memoirs, Cajal mentioned Silveria's skittishness—she is "¡Aquella preciosa niña asustadiza, en que apenas reparé por entonces, resultó, andando el tiempo, la madre de mis hijos!" ["That pretty, fearful child who I then barely noticed but who was to become, later on, the mother of my children!"] (*Recuerdos* I, ch. XIV)—and she remained a figure of similar attributes even after she does become his wife. Settled, earnest, caring, she looked the part she had been chosen to play. The photographs are material proof that she accepted what he wrote of as her personifying "*el ideal de vida perseguido por el esposo*" ["the ideal life pursued by her husband"] (cited in Laín Entralgo and Albarracín 63). He had found a soul mate: "la psicología de mi novia, que resultó ser, según yo deseaba, complementaria de la mía" ["the psychology of my fiancée turned out, just as I wished, to complement my own"] (*Recuerdos* I, ch. XXVII). We have knowledge of this family through the numerous albums and collections of their images, collected to parallel scientific compendia and an activity to test the power of the lens. One of the few photos that shows a calm Silveria facing straight ahead, looking at the camera, is a studio portrait of the couple taken during their first years in Madrid (see fig. 2.9). The differences between this and other studies are obvious: Cajal is seated, hands clasped together, leaning on one side of a chair while Silveria stands at his side. This is protocol repeated in matrimonial studies of the second half of the nineteenth century and has nothing personal about them in its composition. Women attended to their husbands and stood behind them, literally and figuratively. They are just another husband and wife here, not yet a renowned scientist and his bride. Doña Silveria accompanies her husband solidly and steadfastly as he sits. In her hands she holds a small fan, a prop very common in such photographs as part of social etiquette, a delicate prop of femininity, but one that reveals little about her as an individual. Uncharacteristically, both of them look forward, their eyes gazing at the lens of the camera that is recording their image in a long take. They are dressed formally, in dark clothing, and barely touch one another even as they remain side-by-side.

A second formal portrait is full-length without the fade-to-white at the bottom of the previous photograph. Perhaps not merely younger but also more recently wed, Cajal and Silveria appear once again standing and seated, but her arm rests on the chair back near his shoulder, and his arm appears to reach behind to encircle her waist. They look off-camera to something far afield from the man behind the lens, and are turned slightly in a three-quarter shot. We observe their provincial dress—more patterned and less formal than some later urban portraits. It includes a high-necked dress that has replaced a shawl. A painted wall in the background, wooden floor planks, and a dark border place the couple in a home rather than a studio, slightly less stiffly accompanying one another. As Cajal had observed in his manual, time changes the appearance of human beings and these portraits form a montage of their relationship documented during different places and moments. Are they merely consolation, or are they also mirrors of success, fame, sickness and tribulation, loss, and even broken dreams?

In two images, Silveria's likeness represents both a presence and an absence: she is young, then older, with time filling the space between them. She touches that moment for us once again by being there, but also by looking different from other images of herself: she is the four-fold portrait of Barthes—who she was in Madrid, who she no longer was (in Valencia), who she wishes to be, and who the photographer captures her as. This time, her gaze reaches and engages us, unlike any of the shots of her put together by Cajal himself, except for a very large 1896 shot of six children and their mother (fig. 2.13). In this collective panorama, she appears among them in the shortest-to-the-tallest order and not at the end of the line. Almost an innocent among innocents, Silveria and the children all dress in dark clothing (the eldest, Fe could be Silveria's younger double), and no one dares to look away from the camera except Santiago, whose eyes are fixed on something outside the frame. If photographs were to teach something about the classes that had access to such equipment and the talent to learn to use it, they revealed a combination of pictorialism and documentation that did not stray far from traditional portraiture, and a reliance on technology imported from abroad. For Cajal, photography was but one window on the world (a complement to the microscope), one way of accessing knowledge, one of many techniques to answer questions about what Spain looked like at the advent of modern times and one way to show his heroic struggle to "*lograr que el nombre de su patria fuera apreciado en el mundo entero*" ["achieve the goal of making his homeland valued by the whole world"] (Laín Entralgo and Albarracín 12). The equipment may have been imported, but its use could always be innovative and challenging. Histology was not a field most of the population would even recognize, but photographic technology was at the forefront of social life in modern European nations, from newspapers to the royal court and from studios to battlefronts.

Among his other photographic studies, both city and countryside are represented, including Zaragoza and Valencia, with frequent scientific excursions and gastronomic meetings, and panoramic visions depicting Spain. On the other hand, Cajal was always portrayed with identifiable and unchanging professionalism, amid his instruments and lenses that allowed him to be identified with scientific method and studious precision. Long after he took the doctored 1870 self-portrait (fig. 2.14: *Autorretrato Cajal*) with defiant stance and admittedly retouched musculature to create a more forceful presence for the local bullies, arms crossed and just waiting for someone to challenge him to a physical engagement, the scientist eclipsed the man. The first portrait already indicated a knowledge of techniques to alter the image of a physique evidently too weak to leave as is. From this bodybuilding adolescent, his turn to science produced a different image of the body. As a medical student in Zaragoza, he is now shown gazing up at a human skeleton, leaning on one arm and on his books and notebooks as he contemplated bones and structures. Behind him are floor-to-ceiling bookshelves, volumes filed neatly, both horizontally and vertically, filling the entire upper space of the frame as they might fill his brain with knowledge. Scientific material crowds out anything else. Sheaves of paper cover the only oddly disarranged note in this portrait of a man engaged in critical thinking and scientific discovery. The table covered with a fringed cloth that looks like it belongs more

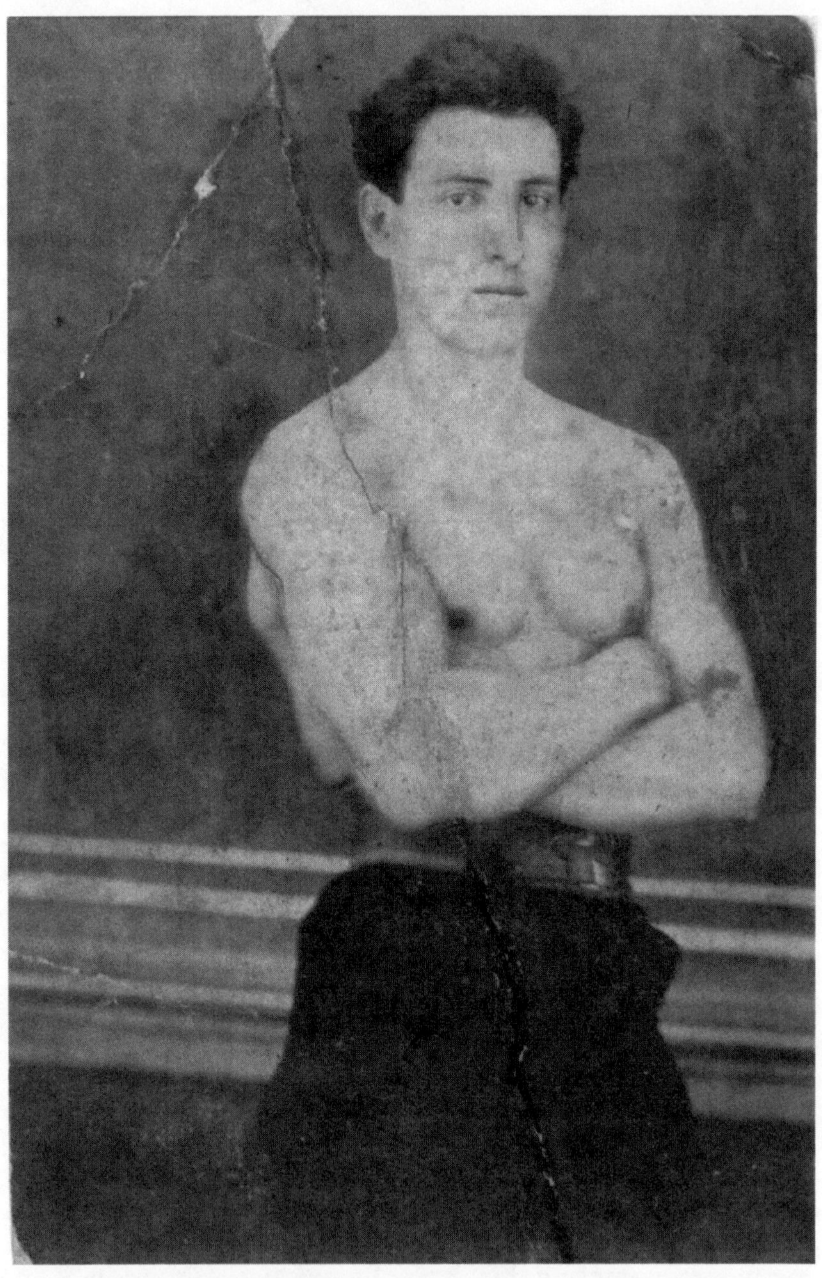

Figure 2.14. *Autorretrato Cajal* [*Self-Portrait Cajal*]. Santiago Ramón y Cajal. Legado Cajal (CSIC). Instituto Cajal. Madrid, Spain. Used by permission.

in a dining room than in a laboratory. This adds a note of incongruity to the atmosphere of science to be created. Cajal is wedged between the skeleton and the draped tablecloth in rapt contemplation of the human body that fascinates him as skeleton, musculature, nervous system, blood vessels, organs of sight, and psychology. There is a silent dialogue between the two, an interrogation of the anatomical structure with scientific investigation as its language.

Cajal photographed himself multiple times in various laboratory settings, from Valencia where he appeared paused in his work with a young collaborator, to Madrid when he had attained fame and public recognition for his drawings and slides. These include both single frame and stereoscopic (multiple, side-by-side) self-portraits. In the 1890s, one photo has Cajal seated at his desk at an angle, both hands on a microtome as if perched in midslice, with a microscope and paperwork covering the surface of the workspace. Half-tones of gray cover much of the scene, except for the contrastive light that falls on his face and on the work surface from what appears to be a window on his right. Other than the studio portraits with doña Silveria, Cajal is never far from his place of work or his confident and steady contemplation of slides and notes. They are part and parcel of who he is. As a rising star in the field of histology, he was not only a *padre de familia* ["father figure; family patriarch"] invested with all of the authority of that role, but a man who stood alone as a celebrity in the public eye. This particular category of portraiture most impacted the Spanish public as photography rose in popularity: the celebrity portrait. In this case, the scientist became a recognizable figure among the politicians, opera singers, writers, or even criminals of the day.

Cajal's 1900 Valencian self-portrait in sharp, detailed black-and-white with the man of science gazing off to the side, not facing the camera, offers a summary vision of one type of artistic portrait grounded in the "solid reality" he sought in his experiments. Here Cajal did not signal a return to photographic pictorialism whose frames were filled with theatrical curtains, elegant drawing rooms, classic architecture re-created out of papier-mâché, or other over-the-top accoutrements to reflect the tastes of society. Yet Cajal did place himself in a specific and identifiable context as far as social class and profession go. A dark curtain covers almost the entire background behind the scientist, with a shelf containing two bottles of dark and light powder emerging on the right side. Cajal's dark jacket, buttoned to the neck, makes his image recede into the background, with only the white collar and cuffs, and his pale hands, to anchor him in material reality. He is serious among the objects of his professional activities. There is a table covered with a patterned cloth that no longer forms part of a domestic scene but has become a surface on which his microscope rests, which might otherwise be a distraction to the eye,. The domestic has turned into the scientific anchored by the tools of research. Powerful, with numerous lenses and eyepieces, this impressive device fills the entire right side of the portrait. The man appears alongside the very visible box of slides because Cajal as photographer has given the equipment equal weight to the man in his self-portrait. Neither is complete without the other. Scientific paraphernalia—evidence of empirical authenticity and material results—are present in this photo in recognizable shapes and repeated "schemes of intelligibility" (Strong 64). Any observer undoubtedly could trace the shape of a microscope after studying the details of these images. To emerge from the evocation of an atmosphere of a kitchen-turned-laboratory, this mise-en-scène has to set the stage to tell, through images, the story of scientific inquiry in all of its various details. But it must do so by means of condensation and brevity, not lengthy narrative or innumerable objects. Photography demanded such synecdoche and metonymy to replace the need for the lengthy description that it replaced. This may be a corollary of Martineau's observation that "the invention of photography shortened the distance between the eye and the hand" (7), reflecting a more immediate and direct link to the visual function of the brain. The microscope singularly challenged the simplicity of a realist vision and substituted for it the promise of what lay under its lenses and beyond immediate sight. It was modernity incarnate.

Perhaps the most challenging self-portraits Cajal produced are a series of microscopic portraits shot in sequence in 1879 or 1880. As part of his frequent experiments with photographic processes and images, Cajal joined the codes of science and art in shots that reduce his face down to the size of what he ordinarily would look at under the lens of a microscope. Rather than seen with ease, the portrait challenged the eye in this new format. This time, Cajal moved from the field of neurons or other structures of the body to the human face as an experiment in visibility. Fernando de Castro, the owner of these miniatures, remarks that the five tiny self-portraits can only be seen using the technology of the laboratory: "Para poder observar esta auténtica 'miniatura,' es necesario el microscopio o una lupa de gran aumento" ["in order to see this authentic miniature, it is necessary to put them under a microscope or a really powerful magnifying glass"] (9). Seen with the naked eye, one only observes five dark spots or dots on a page. They are visible but have no meaning and cannot be read without the aid of an augmenting lens. These images are not revealed as such unless modern technology comes between observer and object, unveiling the unexpected. Cajal made the observer work harder to see. He disrupted the supposed facility of the eye, much as Benjamin proposed technological inventions and innovations forced the inhabitant of the modern world to do. They do not merely recapture the face of the scientist but remind us that seeing is an effort that offers great potential but guarantees nothing. These micro-snapshots are possibly the most representative of the challenge of modernity's culture of the eye for both artists and scientists. The reduction of the human face, after it has been recorded through chemical processes and the power of light, adds not monumentality but transformation to the acts of photographing and describing. One must look hard to see it.

Stacy Hand writes on photomicrography and microphotography as two popular innovations of the latter half of the nineteenth century, as modes of visuality worth exploring to add new dimensions to the growing cult of seeing and being seen (925). There is no doubt that everything Cajal experimented with is not accidental but that it contributed to such an image of modernity in his professional life. Each activity supported his desire to see, and to be a man seen by others, much as he wished Spain to be visible on the world scene for its capacity to contribute to scientific investigation. That the focus of his life was just that—modern science—was a fact that made him different from the average man and, therefore, worthy of spectators. The "curiosity" and stubbornness noted by all as the hallmark of this empiricist have been borne out in the self-portrait taken almost fifty years after his birth. We see a mature, serious scholar fixed for an instant in a pose of great intensity that reflects his attitude toward the two activities to which he has dedicated his time, science and art. His fascination with the lens as an aid to the examination and explanation of the nervous system, as well as to the collection of images to make a long-lasting and permanent record of the faces of those around him, is also seen through the "window" of the microscope that allows the miniature portraits to be revealed.

On the other hand, when Cajal peered out from behind his family in a 1906 collective portrait—fig. 2.13, *Familia*—his face was almost eclipsed by the adult children who surrounded him. His two hands lightly grasping the shoulders of Paula and Silveria tell us that he is the patriarch of this familial group, but his is not the first countenance we see. Instead of a larger-than-life figure occupying the foreground, he emerges from behind the back row, the progenitor but not the family's most visibly imposing member. Cajal was the human foundation, the bedrock, on which the family was built, just as he stood in for the life of science that had won him a public reputation. Even amid the shadows, his hand is not only on their shoulders but he has had a hand literally in everything they have done. He had been successful in his scientific endeavors, as their elegant and modern formal clothing makes known.

His son Santiago is a grown man in his late twenties, kneeling on a velvet cushion in front of his mother, framed by all the others; the sisters occupy a preponderance of the space at the center of the shot, and women outnumber men. Even though the image of the scientist ordinarily dominates the portraits he produced, whether alone or accompanied, that "copious library of images" he described in his memoir was comprised of other scenes such as this as well. He was not the photographer this time around, and the entire family becomes instead the focus of the portrait. Here the venerable father and histologist was obviously present, but only as one component of a greater image, even though perhaps it is the fact that it is Cajal's family that motivates the photograph to begin with.

With this collective portrait, he supplemented his scientific research and collections with data from a cultural project much greater than the mere sum of its individual artistic and scientific components. The album is a timeline of a life amid others: "la serie cronológica de fotografías de un sujeto parece realizar el ensueño de la reversibilidad de la vida, del cinematógrafo al revés" ["the chronological series of photographs of an individual (that) seems to fulfill the dream of reversing life, of the film running in reverse"] (cited in Laín Entralgo and Albarracín 168). Shortly before he died, Cajal's son—and the one in the family he seemed to find most like himself in temperament and intellect—was made materially present. As a scientist his work with matters of life and death made it abundantly clear that there was no reversibility: time flows in one direction only. Cajal uncharacteristically waxed a bit emotional when he referred to a "dream" of making the dead live on, but he wrote this passage after Santiago had succumbed to his illness. Yet Cajal did reveal another dimension to his thoughts when he likened the camera to the cinematic camera, a technical invention capable of bringing images to life in reverse chronology. Of course, that was a technical observation and not a historical process. He seemed very aware of the possibilities of such mechanical reproduction to both bring invisible things to light for the scientist, and to allow the artist to create different visions of the world. Cajal's aspiration to promote and encourage the practice of photography in Spain was therefore both scientific and personal: the machinery of right depiction for the laboratory was also a link with his family and with the past.

In one of his last published papers, in 1918 Cajal included five prints and twenty-two photographs related to the microphotography of the human nervous system among his experiments and conclusions. This brought microphotography and photography together under a single rubric as empirical observation. Seemingly aimed at addressing innovations in unrelated fields, the technologies involved in reproducing scientific information as well as images of other objects are united in Cajal's engagement with experimental photography. Somewhat like Achim proposes regarding the artifacts of the early modern colonial sciences, cultural "truths" are embedded in a variety of often-unpredictable scientific activities (113), and not just where early modern eyes expected to find them. The challenge is to pry open the activities of a given historical moment to find the myriad ways in which the creation of knowledge can cut across fields. Often, technological inventions facilitated inquiry about the natural world as well as the social world, and hinted at emerging interactions between the two. The cult of the lens produced a shift in vision from a reliance on the unassisted eye to the enhanced vision of microscopes and cameras. It also demanded (and produced) a plethora of visible results made accessible to the public as well as to the specialized scientist. Cajal's use of the technologies of the eye presented evidence of a moment in Spanish modernity that echoes European values for both the science and the arts. They preserve an artifact in situ—fixed on slides or glass plates—with all its detail and variation, allowing the observer to hypothesize, test theories, or even abandon previous theoretical conjectures.

Not so much a theory of the family as a theory of new ways to look at and document it, Cajal's ideas about photography supported it as both an art and as a tool at the service of the intense action of scientific culture. The ego of the biologist—evident in Cajal's writings—nourishes his search to explain the formidable enigmas of the world around him. That ego emerged in the laboratory as well as in the family portraits and the self-portraits where the craft and the science of the photograph are put on view. If the photograph sent him afield in search of additional people and places, the histological exemplar resubmerged Cajal in the laboratory where the pursuit of knowledge took over with energy and zeal. Cajal had concluded that faced with what seemed hostile nature, the only response can be to deploy any technologies available to explain it.

In the end, through experiments in hypnosis and in psychology, as well as the investigation of chemical processes in photography and the "science and art of histology" (Triarhou and Vivas 87)—that is, the research as well as the sketches and diagrams—Cajal's discourse on the language of social relations, made evident through portraits and self-portraits, met and complemented the more austere language of the scientist. His work on the retinas of bees and other species carried into the language of the poet and writer in the form of both structural associations and metaphorical ones. The imagination—for him, a fascinating function of the brain—crystallized into signs and symbols (language and images) as the "flow of ideas becomes molded into symbols; it surfaces either in the form of language and gesture or through pen and pencil. Thus, the poet who through his writing or recitation evokes almost all his registers of solemn, painful, or emotive representations, feels, at the conclusion of the work and the restoration of his strength, that his mental retina is imperceptibly tinted in the complementary colors" (Triarhou and Vivas 84). The rededication of the self in the laboratory, then outside in nature, and last behind the lens of the camera, left an indelible image in the brain and on paper. It also established a widening paradigm for the scientific persona emergent in modern Spain. In light of Miguel de Unamuno's renowned rejection of the theoretical sciences in Spain as something "foreign" to the national spirit, having a blindness toward the value of European science in general, Cajal's work with cells and with lenses became critical to a counterargument. The much-cited line "*que inventen ellos*" ["let others invent"], a cliché torn from the intellectual debates of the first quarter of the twentieth century in Spain, appears in a letter from Unamuno to José Ortega y Gasset in 1906. In a dialogue between the characters Román y Sabino en *El pórtico del templo* (also 1906), and the epilogue of *Del sentimiento trágico de la vida* (1912), Unamuno widens the rift between modernity and Spanish culture in contradiction to Cajal's exemplary bridging of the arts and sciences, and Spain and Europe. With Cajal, both a private and a public persona were created that went against such a rejection of the value of scientific thought as alien to Spanish culture. Instead, Cajal as a meritorious model for the man of science looked right at home in the space of the family, the laboratory, and the natural landscape.

Chapter Three

Matter, Time, Landscape

Ways of Seeing in Cajal, Ortega, and Benjamin

The epoch of empirical observation related to imperial Spain closed after the Spanish-American War of 1898. The first two decades of the twentieth century were characterized by competing examples of technological innovation amid a generally insecure social and economic modernity. The touted keys to modernization—industrialization, education, urbanization—were visible but uneven. Modernity as the changed experience of time in a new world where tradition had ceased to hold the same power as in the past was far from universal. The Atlantic world had faded into the background, although empirical practices had been assimilated from the colonies into the peninsula as far as scientific observation had opened the gates to communication, mass media (telegraphy), individual mobility (the apogee of railway systems starting between 1848 and 1866), and had guided popular experience through the collection and dissemination of observed and marketed information.

Instruments such as telescopes and barometers, the topographic Abney levels and macrometers of surveyors, and photographic lenses were interdependent in the observation of the terrain and its inhabitants as well as their potential for future development. Not for the exploration of new territories per se, turn-of-the-century science nevertheless provided measurement and quantification of resources, even as the Spaniards of the provinces often remained at the margins of technology as part of everyday life. The images promoted by photographs could be accurate or utopian, commercial or aesthetic, depending on the location and position of the observer and the desired outcome.

Quite modest a presence in the burgeoning circles of European modernity, Spain would have to cease being the overwhelming traditional recipient of colonial exports and invest in the science and technology that other European nations could offer to bring the society into line with modern dreams and aspirations. Therefore, traces of modernization imported from more advanced countries coexisted with an uneven experience of modernity: the material presence of objects, constructs, and goods was not always connected to the experience of participation in the history and processes to which they might belong. As Spain "struggled to maintain its place as an international power" (Alberto Elena and Javier Ordóñez 70), cities emerged as the sites of cultural conflict. What Benjamin approached as an emblematic "architecture of modernity" (Benjamin and Rice 3), an extremely complex conjunction of elements and readings of urban space, required a liaison with the inventories of particular cities. For him, this included Paris, Marseilles, Naples, and Moscow. His close relationships with those urban centers then fed into more general taxonomies of modernity. An ideal way to observe the case of Benjamin's readings of cities is through the commentaries of Buck-Morss and Habermas, respectively, on his linguistic deficiencies in Russian turning him toward the visual to "see the presence of the Russian Revolution" (28), and on the "unfinished project of modernity" ("Modernity: An Unfinished Project" 38) incorporated into the stones, ruins, and monuments of each city. The same holds true for observing Madrid in the early 1900s, as it held the distant promises of the past embedded in its crumbling walls and the renewed hope for the future in modernist architecture. The completion or near-completion of key buildings such as the Edificio Metrópolis (1901–11) on the Gran Vía in Madrid; the construction of reinforced concrete facades for modernist-inspired new movie theaters at the center of the city; the planning of neighborhoods such as Colonia El Viso in Chamartín, later realized by architect Rafael Bergamín, all attest to the general impetus toward the eye as the organ that could convince Spaniards of their shared participation in a European modernity. If they lived amid this architectural evidence, that would be proof enough. The mass media would contribute to disseminating the images of the architecture that would house communication and entertainment technologies (for example, the *Telefónica* building, the first skyscraper in Europe, built between 1926–29), planned residences, and proof of a modern Spain oriented around such projects. A café habitué, Cajal's

own participation in the life of the streets—and in the regular political, artistic, and literary social gatherings called *tertulias*—reveals his
acquaintance with the power of the lens and the image to sway public
opinion about the city. In the collection of short pieces *El mundo visto
a los ochenta años* (*The World Vision of an Octogenarian*), written in 1932,
Cajal composed a brief entry on photography where he distinguished
between the need for objectivity in the reportage of events and phenomena ("la fotografía documental" or "documentary photography"),
and the "fotógrafo de gabinete" ("portrait photographer") (187) who
finds himself retouching imperfect anatomies of his subjects in order
to please them and to fit into the aesthetic norms of the day. He finds
artifice—the use of retouching the image and layering on makeup to
hide the effects of the passage of time on a face—totally unaccept-
able for documenting news events or the construction of a building.
Yet the lionizing of modern architecture performs a similar function:
razing old structures for the development of showplaces of progress is
equivalent to making crow's feet disappear behind a facade of makeup.
While Cajal falls into a condemnation of modern art and commercial
photography, "todo se ha sacrificado a la comodidad y baratura del
trabajo" ["everything has been sacrificed to the ease and cheapening
of work"] ("La fotografía," 158), he still retains an emphasis on the
continued importance of the utopian accuracy of the images and the
use to which they are put. In fact, the rejuvenation of an aged countenance he describes could be applied just as well to the face of the
city: "el fotógrafo de hoy retoca furiosamente, resta muchos años de
la edad a los modelos y procede, en fin, como los cirujanos llamados
profesores de belleza" ["today's photographers retouch furiously, taking
years off the age of models, following the actions of surgeons now
called professors of beauty"] ("La fotografía," 187).

The deification—and, through the promotion of associated
products, reification—of progress as *the* irrefutable sign of modernity
required the observation of referents to that forward motion, whether
they be drawn from humanistic fields of endeavor or from scientific
ones, from trained researchers or everyday visions. What Holton calls
"the interpenetration [of the arts and sciences] reveals itself through
the use by scientists of metaphors and of the thematic imagination, in
many cases enhanced by literature and philosophy" (133). Music, the
cinema, physics, art, the laboratory, or the open road would coalesce
into a multifaceted series of metaphors for modernity's impact on

Spanish culture; what was needed, above all, was acuity as to what to look for and where to look.

In his *Meditación de la técnica y otros ensayos sobre ciencia y filosofía* [*Thoughts on Technology and Other Essays on Science and Philosophy*], Ortega writes that everything within human sight is part of the dialectic between human beings and their environment that involves the satisfaction of needs as well as the fulfillment of desires. In Ortega's words, "Naturaleza no significa aquí sino lo que rodea al hombre, la circunstancia" ["Nature here means no less than all that surrounds man, his circumstances"] ("Primera escaramuza con el tema," 25). Technologies could reformulate that dynamic relationship between individual and surroundings, exercising power over the environment and distinguishing with a clear eye—"*con una pupila*"—what the priorities of the individual are rather than life merely copying the wants of others, what Ortega refers to as "*una manera bizca de exisitir*" ["a cross-eyed existence"] ("Vicisitudes de las ciencias," 136). Ortega's visual metaphors are evident in many of his essays, but they dominate those addressing the modern concerns of new scientific theories, technology, physics, Einstein, and "los prodigios del presente" ["present-day marvels"] ("El tecnicismo moderno," 91).

In addition to commenting on the relationship of the observation of phenomena to the biological theories of the early twentieth century, Ortega proposed that to avoid a descent into whim and caprice outside all historical context, it was critical to take a multigenerational, longer view of contemporary debates in terms of "un clima intelectual, el predominio de ciertos principios atmosféricos" ["an intellectual climate, the predominance of certain atmospheric principles"] ("A 'Geometrías no Euclidianas,' de Roberto Bonola," 160) that either favored or distrusted certain methods and conclusions. Cajal might have been in agreement about the experience of an era had he discussed this with Ortega, since at eighty years of age the histologist could keep a trained eye on the methods and goals of commerce that did not square with his own. Toward the conclusion of his essay on photography, Cajal writes of the "métodos modernos de fotocopia (proceder de Meisenbach y otros)" ["modern photocopy methods (originating with Meisenbach and others)"] as the media for the preservation of politicians and superstars of the bullring for the eyes of posterity. But just as he lapses into parenthetical comments on the scientists responsible for these modern inventions, Cajal retreats into a more reflexive stance: "Pero entramos en consideraciones demasiado técnicas y aburridas para el lector profano. Además, nadie debe hablar demasiado de lo que sabe; porque ello produce tanto placer al que escribe como disgusto al que lee" ["But with that we are entering into too many technical and boring considerations for the profane reader. Besides, no one should talk too much about what he knows; that produces as much pleasure for the writer as it does displeasure for the reader"] (159). With these words, he appears to once again acknowledge the two readerships of the time—a scientific one and a popular one—and that each expects to read language and images differently. The need for artifice and makeup corresponds not to science but to commerce and to the values of youth, modernity, and rapid change.

Ortega is not alone in turning to literary metaphors to capture the often-jarring juxtapositions of the new and old, modern and traditional. After all, this figure of speech binds together the known with the unfamiliar, comparing what may be recognizable with something foreign, creating an analogy by borrowing attributes and ascribing them to another, just-produced object. Capitalism runs with speed; abundance is evidence of progress, as what Buck-Morss terms dynamic "semantic constellations" (*The Dialectics of Seeing* 91) are born. Both Ortega and Benjamin uncover the armature of modernity to search for survivals, remnants, for that flash of awakening from a passive intoxication with the new. Landscapes, both urban and rural, were the stages set for stunning and evocative visual encounters with past and present.

In an essay from 1926, Ortega starts his journey among Iberian people and lands with the *ermitas* in Córdoba, those small sanctuaries far from towns, to trace the silence of the past in the face of the noise of modernity. Calling these chapels "*sanatorios de silencio*" and "fábricas de soledad" ["sanatoriums of silence and factories of solitude"] ("Las ermitas de Córdoba," 13), Ortega evokes the sense of time's passage through the description of space. The last word of the essay, *cal*, the lime used to cement stones into buildings, is both literal and allegorical. The element that bound the sanctuaries together in faith was also the material that still stands as a monument to something now lost, or at least in the process of disappearing. From there, he remembers the fountains of Nuremberg (their waters singing a paean to the past, while everything else around them changes); Castillian landscapes, sunsets in Asturias, China, summer in Basque country; the hidden cultural messages in clothes hung out to dry; and a painterly discussion of Spanish artists' seeing beneath the surface of things to bring objects to life. Influenced by the art school of Venice, Ortega finds Spanish painters more tactile than visual, "diríase que el pintor los ha mirado [a los objetos] con las yemas de los dedos" [one could say that the painter has seen the objects with the tips of his fingers"] ("De Madrid a Asturias, o los dos paisajes," 88).

At a new distance as well as at a new speed, Ortega can find the "porosidad" ["porosity"] (89) of the landscape penetrating the sky: light and rock fuse, things lose their definitive outlines and the concrete evaporates into color and light. The rapid circulation of people and commerce in modernity shifts the tactile to the visual, as contact is fleeting. The advent of the photograph in place of the hand touching the actual stone removes the intimacy of experience between people and things, leaving behind instead a decay of captured tradition (in the image) as uninhabited ruins. Ortega's eye revives the landscape of ruin with the reenchantment of descriptive language, fills the silence of abandonment with a new voice: "donde laten las entrañas de las cosas, . . . esperamos que rompa a hablarnos cuanto no sabe hablar" ["where the hearts of things throb with life, . . . we wait for everything that knows not how to speak to talk to us"] ("Temas del Escorial," 47). The first step, of course, is to observe these places and objects for them to become animated with life. While Cajal would most likely categorize these images as painterly and not objective, filled as they are with evocative language and philosophical comments, they are evidence for the empirical proof of a Spain both material and spiritual and are no less imbued with emblematic meaning than the stones, walls, and accumulated lint of factories that Benjamin examines. The portraits of family members and walks in the country that Cajal produced are equally emblematic of lost time. What he adds is an aside about the processes used to record the images, not just the photographs themselves.

In search of illumination, of knowledge hidden in the objective world, Benjamin seeks out "the small, discarded objects, the outdated buildings and fashions, . . . the 'trash' of history, . . . the evidence of its material destruction" (Buck-Morss 93), the small and the fragmentary, as an antidote to the "elephantism" of the state's power (92). Cajal could figuratively make his deceased son come alive by contemplating a photograph and comparing it to the image in his mind's eye. Benjamin traversed the "hollow and crumbling shell of the precapitalist order" of

Naples (Buck–Morss 26) in a similar attempt to revive the stones into speaking. Both are traces of the past, one physically material and one a representation of a moment, but each can spark that almost chemical flash of revelation. Cajal, Ortega, and Benjamin all deal, each in his own way, with the rise of the masses, mass culture, through the rereading of observed phenomena through a critical lens. Counter to Madrid, Ortega travels the provinces; counter to the elaborate constructs of commodified modernity, Benjamin inhabits the alleys, cafés, ruined walls, and other traces of the past in Naples, Moscow, or Marseilles.

Between the end of 1926 and the first months of 1927, as Ortega contemplated the crumbling panorama of provincial Spain, Benjamin traveled to Moscow to see for himself the images of the Revolution. What he observed among the people and their everyday activities became a new language of images, not words. In a letter to Martin Buber on his return, Benjamin explains what he has set out to do in his diary: "I hope to succeed in allowing the 'creatural' to speak for itself" (*Moscow Diary* 132). Benjamin, too, resorts to metaphors to communicate the look of the streets: begging is "a corporation of the dying" (*Moscow Diary* 105); in comparison, the streets of Berlin are an echo of desolate scenes by artist George Groz (*Moscow Diary* 97). "Each thought, each day, each life lies here as on a laboratory table" (*Moscow Diary* 106). The face of the city houses both new and old mythologies, traditional culture and modern technology, reality and appearance, fashion and boredom, those who kill time and those who live accelerated lives. As Michael Ugarte recapitulates Benjamin's work in the Arcades project, "perhaps in the last analysis, Benjamin's reading of nineteenth-century Paris represents his urge to lay bare the *concreteness* of culture, its materiality both in relation to the commodity system and to the social practices of which it is both the product and the producer" (15). Cajal, Ortega, and Benjamin all observe the urban spaces of modernizing cities that reveal paradoxes and contradictions, their ideas and observations based on the most visible, material evidence.

The concern of Cajal, Ortega, or Benjamin was not the preservation of an archetype but a juxtaposition of alternatives in the scientific, cultural, or historical landscape of change. Decisions regarding topics, perspectives, perception, and the use of illustration as a strategy to convince were based on the effectiveness of the information provided. The emergent reception of photography as an authoritative visual medium, yet also one that could be used for creativity, was critical to a considered rereading of what the unassisted eye could perceive and what the camera lens recorded.

The term *landscape* is generally used to refer to certain perceptual aspects of the natural world, such as geological formations, ecological systems, geographical attributes, or, when citing the details of urban horizons, architectural structures. It can also encompass all of the visible features of a specific area, the human inhabitants, and their constructs. In the case of modern architectural landscapes, Benjamin noted a shift in the perception of them as a move from the "tectonic" to the "optical" as part of architecture's "new status as media" (Leach 19) or mediator between humans and their environment. In other words, just as the eye was supplemented and then supplanted by the lens in the laboratory, the camera in the dark room, and the cinematic apparatus in the movies, architectural forms "shift from object to image and from the tectonic to the surface effects" (Hartoonian 25), with the arrival of new technologies that took the place of less complex material relationships. The interrelationships between observer, mediated constructs and their surfaces, and the vegetation, subsoil, climate, rocks, and fauna of the natural landscape merge to form both objective and subjective perceptions of those interfaces between people and the world. The cultural landscape embodies how humans have dealt with those surroundings beyond the first step of perception, and images that capture such features freeze them, distance the observer from them in order to promote more careful study. As Prodger has noted in the case of Darwin, "provide representations that could be used to prompt comments from others" (15). In an era of the reevaluation of previous methods of representation, the camera elicited for all observers a confrontation with what was depicted.

It could be inferred that any attempt at depiction—whether it be linguistic or visual—was thoroughly analyzed and held up to scrutiny by scientists as well as intellectuals in general. José Ortega y Gasset's travel writings as an observer from the rail car or automobile, Manuel de Terán's anthropological record of the evolving cultural geographies of Spain, or Cajal's urban portraits all referenced moments in time that led to meditation, repetition, analysis, or comparison. The bodies of work produced showed the public how to look at images, how to read landscapes in cultural terms, how to insert images into temporal and spatial contexts, and what to expect (or not) from them. How mimetic (realistic, authentic, genuine) they were, their power to convince that the original observer actually "saw" what he wrote of or photographed would render the product either an illustration or an illusion and the intellectual from whom it came a potential figure of authority. Ortega went so far as to write of the absolute need for observation to put a priori conjecture to the test—to use the "*pupila*" ["pupil of the eye"] (Ortega, "Vicisitudes en las ciencias, 1930," 136) in order to avoid falling into imprecise "*cegueras*" ["sorts of blindness"] ("Vicisitudes en las ciencias, 1930," 141) or faulty sight like those who keep one eye on their own life and the other on those nearby, creating a "*bizqueo*" ("Vicisitudes en las ciencias, 1930" 136) or cross-eyed perception of what they thought they saw accurately.

In the case of Spain, scientific practices and innovations that emerged from the activities of engineers, biologists, doctors, and others were produced in large part in the modernizing metropolis that served as a stage for change. The city itself grew in particular ways in accordance with how scientists and intellectuals viewed the need to weave together human actors and the landscape. As Antonio Lafuente and Tiago Saraiva explore, "Madrid . . . became an experimental laboratory in which machines and experts objectivized problems, gathered data and drew up plans of action. . . . it was not just a work place; it was also a patient prostrated on the operating table" (531). After Barcelona and Zaragoza in the case of Cajal, after Leipzig, Berlin, and Marburg in the case of Ortega y Gasset, the Spanish metropolis was the catalyst that brought systems of thought together, and expanded the horizons of science as the city itself grew. Walter Benjamin's understanding of Paris, Berlin, Capri, Marseilles, Naples, and other European capitals articulates a biting commentary on urban experience and cultural perception that unites Cajal and Ortega under the sign of the fragmentary "thought-image" or *Denkbilder*. These brief city portraits that constellate into cityscapes captured in fragmentary images "seek to capture the fluid and fleeting character of metropolitan existence" beyond the mere banality of the tourist vision. Instead, the essays of the philosopher, alongside the recollections and photographs of the scientist, "dissect with the keen critical eye of the physiognomist" (Gilloch, *Walter Benjamin: Critical Constellations* 93) the landscapes of the early twentieth century as laboratories of modernity. The intimacy of the pedestrian, added to the access of the streetcar and the train, accord a physical proximity between the spectator and the profusion of objects available to the eye (and the wallet) in the jostling crowds of the city just as these modes of travel open panoramas for scrutiny in the vast regions of the countryside.

The extension of the city into formerly rural areas, the dispersion of population into greater circles of habitational space, the expansion of scientific institutions all linked landscape (urbanscape, cityscape) in a "drift towards the city outskirts" (Lafuente and Saraiva 536). New sites such as the train station of Atocha and several other axes formed around a new city plan. City and science intersected in their development, creating buildings and the subsequent need for transportation and communication between them. The best of the town and the best of the country met through interconnected rail lines, reservoir systems, hospitals, University City, athenaeums, and social institutions. On the one hand, scientific activities can link theory and practice, abstractions and their physical implementation; on the other, a scientific vision fragments objects into components able to be analyzed and—potentially—understood. So city life, as well as the vestiges of the life in the provinces many had left behind, became part of public discourse for theoretical scientists. In general terms, life was examined piece by piece before a synthesis might be reached. Between the 1850s and the 1920s, "the old hegemony of botanists, astronomers, and architects began to decline, giving way to new players, chief among them were the professors, the doctor-surgeons, and the engineers" (Lafuente and Saraiva 533). These new professionals met in public spaces, where dialogue and debate flourished and the cultural role of the court no longer held.

Juan José Ibáñez proposes that Spanish philosopher José "Ortega y Gasset utiliza el paisaje como una expresión del carácter de los pueblos" ["Ortega y Gasset uses landscape as the expression of the character of a people"] (1). Two points of view merge into one: the personal and the scientific. The reaction to what is viewed brings observer and objects into contact so that they may influence each other as previous experiences are brought back into play through memory. The analytical perspective of the scientist is not antagonistic to this. The optic of the researcher and the studied contemplation of the observer unite in what Benjamin proposed as the "porosity" of surfaces brought in contact through the eye. Porosity "as a temporal concept" is an interpenetration of subject and observed object that "undoes" (Andrew Benjamin, "Porosity at the Edge" 43) fixed time and opens up singularity into multiples. Organic chemistry, biology, and medicine share the attribute of porosity as it refers to the small openings that allow organic bodies to pass through and penetrate one another. The photograph and architecture are porous as they invite the mobile eye to cross into unsuspecting territories, surfaces, lines, and spaces.

Walter Benjamin and Asja Lacis modeled their 1925 essay on the traveler's perception of the architecture of the city of Naples on this general concept of porosity that reaches into the depths behind the surface. It also has echoes in the travel writings of Ortega and the landscape photographs of Cajal. Buildings, courtyards, stairways, empty spaces, and crumbling walls are the stages for new and dramatic encounters that revive historical connections and personal emotions or evoke new sensations lost in the "habitual, distracted" (Leach 19) apprehension of surroundings in the modern age. How images are gathered, as well as the portrait formed from their combination, articulated both the method and the practice of gathering those observed fragments. Benjamin's concept of porosity fills his studies of modern cityscapes, his term referring to "a lack of clear boundaries between phenomena, a permeation of one thing by another, a merger of, for example, old and new, public and private, sacred and profane" (Gilloch, *Myth and Metropolis* 25) all of which are deployed to indicate time and impermanence, daily life as improvisation, and social and architectural instability. The crumbling of old buildings, added to the framework of new constructs, forms a skyline or urban scape of incompletion and indeterminate time, except for the relative passage and temporal actuality of the wandering observer. Cajal, Ortega, and Benjamin shared a dedication to the observation and representation—in words or images—of traveling through the radically changing landscapes of city and country.

María del Carmen Paredes Martín has studied the concept of *"paisaje"* or landscape in the essays of Ortega y Gasset and concludes that while certainly a recurring theme, it is not a static one but a "noción más o menos compleja, determinada o difusa según los contextos [y] un referente que contribuye a organizar en torno a sí ideas y reflexiones sobre la circunstancia humana y el trasfondo de la misma" ["more or less complex notion, precise or vague according to the specific context (and) a referent whose prime contribution is to allow the coherence around it of ideas and reflections about the human circumstance and its backdrop"] (177). Ortega's observations on *paisaje*—landscape—as opposed to *naturaleza*—nature, that trope so characteristic of the previous intellectuals of the Generation of 1898—shifted the focus from "*lo inerte y ajeno al hombre*" ["what is inert and strange, foreign, outside humans"] (Paredes Martín 181) to dynamic interaction with an environment that is transformed by point of view from a metaphorical *naturaleza muerta* into a spirited *naturaleza viva*. With this, rather than scientific paradigms of the subsoil or the architecture, observers break and expand the perception of landscapes into multiple allusions and a plurality of fragments producing constellations that form the vibrant and vital "*contorno*" ["surroundings"] of each human being.

Delving beyond the architectural surfaces and the visible landscapes, Ortega himself found in the traces of the visible remnants and vestiges of something more. In 1929, just four years after Walter Benjamin and Asja Lacis publish their seminal essay on the porosity of the city using Naples as their source, Ortega visited the concept of the study of landscape and mentioned the utilitarian aspects, the evocative aspects, the tectonics and the architectonics. He concluded that

> El paisaje posibilita "sentir," "amar," "sufrir," "vivir" en su interior hasta el punto de fundirse con él o quedar sumido en su entramado; invita por consiguiente a trascender el mero estar 'en medio' de él, o 'frente' a él, o simplemente 'ante' él como un observador o espectador habitual . . . Entonces el paisaje se hace repertorio del drama de la vida, se integra en el bagaje de los recuerdos."

> ["Landscape makes feeling, loving, suffering, and living within its spaces possible to the point of becoming one with it or of merging into its structure; therefore it invites one to transcend the just being 'in the middle' of it, or 'in front' of it, or simply being 'in its presence' as a habitual observer or spectator . . . Then landscape becomes part of the repertoire of life's drama, it gets integrated into the baggage of our memories."] (cited in Paredes Martín 177–78)

Benjamin's "dramatic performance" (Gilloch, *Myth and Metropolis* 25) of human actors on the stage of the urban landscape is similar to the "*conjunto de relaciones*" ["totality of relationships"] (Paredes Martín 182) Ortega evokes between subjects and surroundings that are the product of changing and kinetic reevaluations through the eye: "la perspectiva visual [que] nos brinda el aspecto concreto e individual de un determinado paisaje" ["visual perspective provides us the concrete and individual aspect of a specific landscape"] (Paredes Martín 185). The material quality of that encounter is recorded in the essays of Ortega and the *Denkbilder* of Benjamin, while for Cajal the images appear in his albums and collections of photographs. Manuel de Terán's work in the provinces and the urban centers of the peninsula yielded maps, images, landscape sketches, and analytical reports of his observations in the field.

The landscapes of modernity both presented Spanish culture with a reorganized hierarchy of the value of visual perception and offered the technologies that might support and enhance that radical shift in attitude toward what Jürgen Habermas has called "the very 'table' or 'archive' of . . . knowledge" ("Modernity: An Unfinished Project" 38). The very sense of recognition associated with one's gaze on an object was being interrupted—if ultimately also enhanced—by the mediation of new technologies that put into question what had been assumed. That modern disposition, as Habermas continues, "again and again expresses the consciousness of an epoch that relates itself to the past of antiquity in order to view itself as the result of a transition from the old to the new" ("Modernity: An Unfinished Project" 38). Rather than mourn the loss of some codified experience, modernity signaled a crisis of historical representation that demanded engagement with perception in place of the complacency of the habitual or distracted gaze. Ortega placed this *"terrible desconcierto de la vida europea"* ["terrible chaos and disorder of European life"] (Ortega y Gasset, "¿Qué es la técnica?" 14, 14) a result of the rift between intellectuals and technologies. *"Técnica"* as the way in which human beings modify or otherwise mold nature to fulfill their needs produces new landscapes at every turn, he concluded.

Through a so-called crisis of the object, Walter Benjamin expressed that the nature of the cultural, ethical, and epistemological shift in ways of seeing was a reference to the "complete loss of a direct apprehension of the past" (Leach 9) and a turn toward new—if now mediated by the technologies of the eye—knowledge. Such mediation took place through the lens: of the cinema, of the camera, of the microscope, of the retina of each observer. The material artifacts so apprehended entangled both themselves and the technologies in new relationships that scientists, artists, and intellectuals in general had to address owing to the demand to "rewrite and re-image the guidelines used to divide nature into its fundamental objects" (Daston and Galison 16). Ortega's travels through Spain and abroad; Cajal's to Italy, Germany, the United States, and throughout Spain; and Benjamin's flight across Europe all produce texts that reflect the cult of the eye.

In order to read the photographic and linguistic products, which represented, and often scrutinized, the epistemological shift in experience and values that formed a bridge from the nineteenth into the twentieth century, the task of modernizing Spain by incorporating "science" into a more encompassing "climate of civil discourse" (Glick, *Einstein in Spain* 9) must constitute one significant part of how to articulate that world. A social pact to create a social space in which the creation of modern educational institutions was facilitated also allowed for the rise of the scientist as a figure of import. Albert Einstein became a superstar immediately upon his arrival on the scene in 1923. While certain areas of science were represented more strongly, there were small groups of scientists housed in several key university settings—among them Madrid, Barcelona, and Zaragoza—who were cohorts of Cajal in biology, medicine, pharmacology, and public health (although these are very few in number as he recalls in his memoirs). Historian Thomas Glick concludes that such a political "consensus emerged around 1900, after a full quarter-century of agitation on behalf of the scientific ethos" (*Einstein in Spain* 15), and after a national consciousness-searching following Spain's defeat in the Spanish-American War in 1898, in a remote empire. Darwin, evolution, and the 1859 publication of *Origin of Species* had nurtured incipient debates in the 1850s and 1860s, but toward the turn of the twentieth century two buzzwords filled Spain's cultural discourse: *modernity* and *science*. The steadfast way to the first was through a dedicated cultivation of the second.

Although the position of science in general in 1900—the year that the Ministerio de Instrucción Pública y Bellas Artes [Ministry of Public Education and Fine Arts] was created—was tenuous, the last few decades of the nineteenth century had witnessed the formation of the Junta de Ampliación de Estudios e Investigaciones Científicas (JAE) [Council for the Development and Expansion of Scientific Studies and Research] with biomedical sciences, hygiene, toxicology, criminal anthropology, Darwin, and, subsequently, relativity and Albert Einstein as mainstays of what Portela and Soler refer to as "*el flujo de entrada y salida*" ["the flow in and out of the country"] (93) of scientific ideas and figures. Discussions on the topic of Spanish modernity took place in *tertulias*; reports and editorials appeared in many newspapers. Pedro Lain Entralgo and Agustín Albarracin open the door to an examination of the use of Cajal by all political agendas of the time, creating what they refer to as "*una realidad, un mito y un problema*" ["a reality, a myth, and a problem"] (12) in the process of promoting the figure of the scientist as the exemplary future citizen. Cajal was perched at the difficult crossroads of past and future, precariously balanced on a tightrope of social tradition and scientific modernity. Just as Ortega concludes of Einstein's theory that as an historical phenomenon all theories deserve examination in light of their times, so Cajal's work merits consideration as a product of a transitional cultural and scientific moment.

The first of the three terms used by Laín Entralgo and Albarracín—the *reality* of the scientist—refers to Cajal's material investigations into the nervous system; the second—the *myth*—refers to the political uses made of the figure of the scientist to support an ideological agenda; and the third—the *problem*—refers to the ideological battlefield of all those who after the fact claimed rights to the scientist's success. The result is an elevation of Cajal to the status of a hero by some who wished to prove to Europe and the Americas, with great national pride in a culture where the concept of nation was still in contention, the existence of a first-rate Spanish scientist. For others, Cajal functioned as a "*lavaconciencias*" (Laín Entralgo and Albarracín 12), or symbolic moral cleansing for those who never supported or accepted, much less made the path easier for, what scientists like Cajal were doing. Touted in the press as the most universally recognized Spanish scientist for his histological preparations, even without any consideration of his work in the realm of photography, Cajal became a hero in spite of the society around him, as scientific research turned into a collective project of Spanish culture, at least for the nation's intellectual leaders. A third group of Spaniards, usually far less aware of the debates of civic discourse, was oblivious to the rise of Cajal, except as a man whose name appeared sporadically in the press.

Cajal was far from either being left behind in this ferment or kept back by it, since he had been campaigning for the professionalizing of the scientist since the mid-nineteenth-century. Removing the popular notion of the man of science as an outsider, a delver into demonic realms opposed to religious doctrine, or a secretive and dangerous investigator seeking to dethrone the pinnacle of the natural order of the universe would be difficult indeed. Whether in the domain of the officially designated workspace or of the home-converted-into laboratory, or in the rented space of the darkroom, as Cajal would later have in Madrid, it was scientific activities that he promoted—with understatement as his "*modestas contribuciones teórico-experimentales*" ["modest theoretical experimental contributions"] (*Recuerdos* II, ch. II)—as evidence of a subculture existent in Spain that actually flourished. Cajal's work produced strong support and admiration in laboratories from Paris to Berlin, even if many fewer at home. Looking back at the trajectory of his illustrious career, Cajal's reiterated lament that "Debíamos luchar con el prejuicio universal de nuestra incultura y de nuestra radical indiferencia hacia los grandes problemas biológicos" ["We had to fight against the universal prejudice against our lack of culture and our radical indifference towards the great biological problems"] (*Recuerdos* II, ch. II) affirms that such resistance only spurred on his own work in histology, despite such a path leading him only to the faint praise of a small group of knowledgeable men and not the admiration of those doing work in other fields. The real scientist was, he affirmed, not a figure attributed to Spain; whereas the "*poeta melenudo*" ["long-haired poet"] was (*Recuerdos* II, ch. II). Laura Otis perceptively remarks that Cajal "was motivated by an ideological perspective that became a scientific attitude" (64); there was no aspect of the natural or the social world that escaped his critical eye or careful observation and no political force that would keep him from pursuing his interests. He was a man motivated by histology and the mysteries of the cell, to which he devoted himself religiously. His days were filled with tireless labor, stubborn willpower, and a devotion to objectivity as an antidote to emotional interference with scientific experiments he witnessed among some colleagues.

Cajal's minutely detailed reports on procedures for nourishing tissue samples or for developing photographic dry plates revealed him to be a cornerstone of the cult to education, an individual driven to activate and sustain intellectual investigation that would, subsequently and eventually, take Spanish society in the direction of other modern nations. Ortega's education in Germany exposed him to similar theories and ideas that would be the focus of his essays on his return. As always, Cajal was motivated by accuracy in his drawings and, later, in his photographs. He offered others the material evidence of his own inward precision and a good eye. While drawing was a discipline at which he had excelled from an early age, photography was an activity that required practice and discipline to capture on the surface of the world what was underlying its visible structures. Photography then was an art that could come to the aid of science, both the social and natural sciences as the discipline of cultural anthropology arose alongside the "discontinuous transitions" (Caygill 121) in the social world that accompanied modernity. He was a recorder of transitions in many senses, from idealist imaginings of the domestic to skillful representation of the natural world, from looking at landscapes to minutely describing them as part of scientific investigation, and from an emotional account of experience to a scientific account of a world that no longer resisted human understanding but was transformed by it.

Cajal's collection of photographs—among them some of his own and others not attributed as his own work—included a number of panoramic shots of the evolving skyline of Madrid, projecting over the rooftops and into the distance, as well as street scenes. As data, these multiple shots inform the viewer as much about the city as they do about the development of photographic methods. Cajal's street scene in fig. 3.1: "*Escena matritense*" combines the sense of a crowd with sidewalk vendors alongside restaurants, people in rural dress (caps, corduroy, and wool sweaters) next to men in three-piece suits, caped dress coats, bowler hats, and ties. Cafés, parks such as the centrally located El Buen Retiro that had passed from the monarchy into public hands since the mid-nineteenth century, theaters and cinemas, a growing number of photographic studios such as the one Cajal opened along the Castellana, and the changing architecture of the urban environment begin to document a spatial arrangement of the city that was more an agglomeration than a totality. Gilloch regards Benjamin's essay on Neapolitan life as an antidote to Western organized cities and social structures, a paean to the contemporary city's juxtapositions. The monuments to an imperial past—palaces, churches, convents, government buildings—do not appear in Cajal's photos, just as Benjamin's essay captures "brief but vivid descriptions of the city's buildings land streets, markets and festivities, beggars and children" (Gilloch, *Myth and Metropolis* 24). Neither Cajal nor Ortega visited cities or countryside as tourists—and cultural geographers would shift the focus from romantic longings to studied

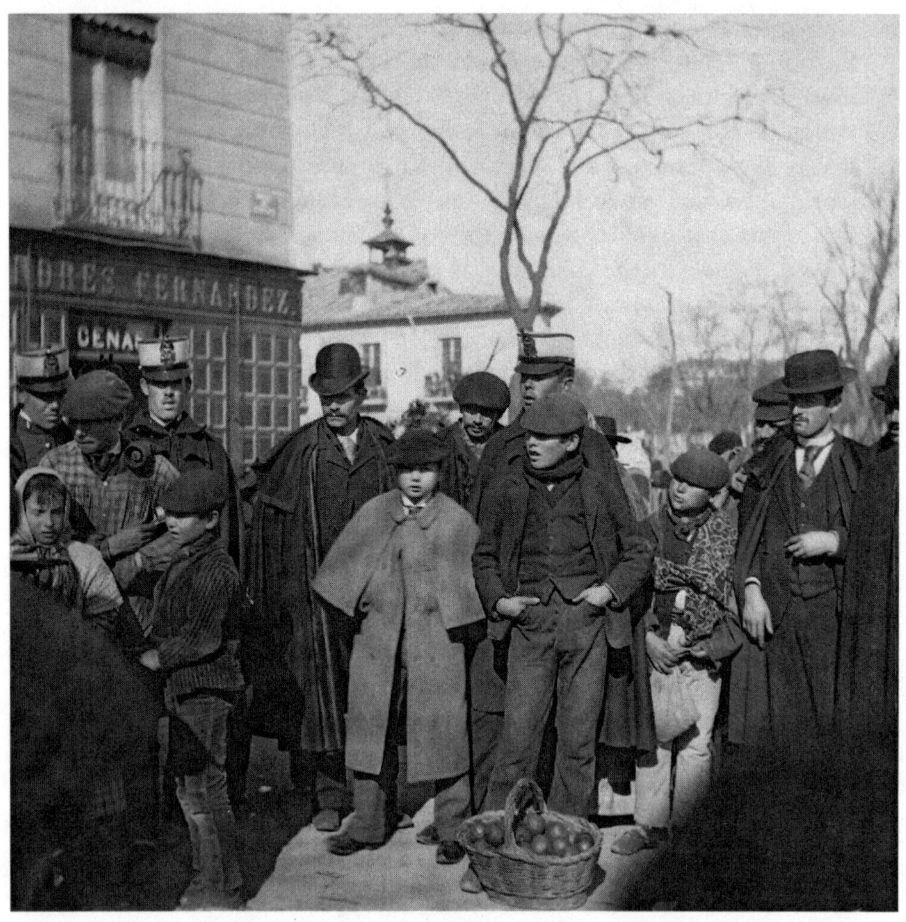

Figure 3.1. *Escena matritense* [*Madrid Scene*]. Santiago Ramón y Cajal. Legado Cajal (CSIC). Instituto Cajal. Madrid, Spain. Used by permission.

hypotheses—but instead as participants in burgeoning capitalist enterprises, with all of the new trappings set among the ruins of the old. They both record facets of the conversion of Madrid from imperial capital to urban metropolis: in boulevards, cinemas, parks, buildings, diminishing greenery, the juxtaposition of old and new architecture, bustling markets, university expansion, café society and *tertulias*, urban migration, the construction of railways and roads, and the emergence of new forms of entertainment such as the phonograph and the movie.

Cajal responded to the growing need for photographic documentation and knowledge, seen as having social and scientific as well as economic value at the time, with the publication of his manual on color photographic processes. He took numerous photos of Spain's varied geographical regions, traveling around the country following his rest from the catastrophic experiences in Cuba. His memory fed curious observation and fused with it with images and language reflecting his convalescence. He took them for himself, as a relaxing pastime or as weekend adventures in various Spanish regions, but could circulate them among others as well as documentation of where narrative might fall short. Cajal could use the images to fill in for times, places, and people outside the new social structures, or no longer appearing in the expected landscape. As Nadir Lahiji comments about the intersection between technology and the production of information, Walter Benjamin theorized that these new forms of media had become indispensable for the production of knowledge, transforming the machinery from a "domination of nature" into an "interplay with nature" (75). The entire experience of observation and recording had changed with the camera. This was slightly different in the laboratory where the scientist was using technology to open up spaces previously invisible to the human eye, but the two complemented one another.

In the case of photographic images of other subjects, experimentation with the enlargement or diminution of images to require careful scrutiny and sometimes the use of lenses, the slowing down of action, or the chance encounter with a landscape or a face now allowed for access to "the space informed by the unconscious," according to Benjamin (cited in Lahiji 80). That is, the image mediates gaps in time and motion, capturing "what prevents sight to be immediate and present" (80). Photography slows time down, it links two separate moments; the processes of photography fill in—or stand in—where the eye misses.

Roland Barthes has pointed out that the special consideration of the photograph is that it offers the viewer an experience that is "truly unprecedented, since it establishes not a consciousness of the *being-there* of the thing (which any copy could provoke) but an awareness of its *having-been-there*. What we have is a new space–time category: spatial immediacy and temporal anteriority, the photograph being an illogical conjunction between the *here-now* and the *there-then*" (Barthes, "Rhetoric of the Image" 44). The image of the photograph is present in front of the observer, but the objects represented by the image are not. As Barthes concludes, the "*real unreality*" of the photograph "is the always stupefying evidence of *this is how it was*" (Barthes, "Rhetoric of the Image" 44), capturing the moment of time past in the shape of the material object past that now stands before us. A conundrum is thus created by the rupture produced through the then/now of the image alongside Martineau's observation that, for the artist and scientist, "the invention of photography shortened the distance between the eye and the hand" (7). Representation gives access to the visual object while at the same time it permits a connection with something that is no longer there.

In his old age, Cajal tempered youthful enthusiasm with a more down-to-earth assessment of the task of the singular researcher. He wrote of the difficult relationships between the great and the small, the cell and the organism, the individual human being and the society:

> . . . en toda nación civilizada la concurrencia vital se extingue o se atenúa en gran parte por la división del trabajo, que hace a los ciudadanos solidarios en sus intereses y aspiraciones, también en el estado orgánico, gracias a la previsión de las células nerviosas y al citado reparto profesional y, en fin, a la supresión del ocio y de la excesiva libertad individual, etc., la lucha desaparece o se dulcifica, se compromete gravemente por causas interiores o exteriores. En otro pasaje hacía notar, en coincidencia con muchos biólogos y filósofos a quienes no había leído, que la naturaleza sólo se preocupa de la vida de la especie. Una existencia, por grande que sea, aun ennoblecida por los fulgores del genio, nada significa a los ojos de la Naturaleza.

> [In every civilized nation vital competition is done away with or greatly attenuated by the division of labor which

gives the citizens common interests and aspirations, so also in the organic state, thanks to the foresight of nerve cells and to their assigned roles, and finally to the suppression of idleness and excessive individual liberty, etc., the struggle disappears or is moderated, seriously compromised by either internal or external causes. Elsewhere I have noted, along with other biologists and philosophers whom I had not read then, that nature is concerned only with the life of the species. A single life, however great it may be, even though ennobled by the fires of genius, signifies nothing in the eyes of Nature. . . ." (*Recuerdos* II, ch. II).

Yet the cell remained his minimal unit for all arenas of consideration, whether literal and scientific, or social metaphors. But his jubilant hopes at seeing the results of the Golgi method of staining and what he might discover with it rescinded any feelings of shortcoming. He concluded:

> Las dos grandes pasiones del hombre de ciencia son el orgullo y el patriotismo. Trabajan, sin duda, por amor a la verdad, pero laboran aún más en pro de su prestigio personal o de la soberanía intelectual de su país. Soldado del espíritu, el investigador defiende a su patria con el microscopio, la balanza, la retorta o el telescopio.

> [The two great passions of the man of science are pride and patriotism. He works, no doubt, from the love of the truth, but he labors still more on behalf of his own prestige or the intellectual supremacy of his country. A solider of the spirit, the researcher defends his native land with the microscope, the scale, the retort, or the telescope.] (*Recuerdos* II, ch. III)

Despite Cajal's suffering in the 1870s, as he learned firsthand of the realities of the Cuban landscape filled with tropical diseases, he did not return to glorify or abstract Spain but to record its changes and the challenges as well as opportunities of capitalist modernity. Even as he mentioned the possible loss of faith in certain methods, he assured readers that his faith could be tested but never lost. Cajal was not a scientist propelled by the economics of fortune but rather by an internal struggle of his own not to abandon those intellectual "fires" within. If he came to the aid of the nation's image on the world scene at the same time, his own intellect—and its strict methods of investigation—would also be reinforced.

The very same qualities that propelled Cajal's remarkable capacity to "alternar sus tareas en el laboratorio y en el cuarto oscuro con la observación al microscopio, la pluma y las cuartillas" ["alternate between his work in the laboratory and in the darkroom with observation under the microscope, the pen, and written pages"] (Fernando de Castro 6) bridged the separation between home and lab, public and private, the linguistic and the visual, art and science. When challenged by spare time after his labors of the laboratory were through, Cajal formed a recreational society dedicated to making Sunday excursions more than the minimal family outings, to exploring regional cuisine, to photographing landscapes with his Kodak camera and, ultimately, to his using photography to perpetuate fleeting images of the past, *"cuajando en pruebas que guardamos piadosamente, como recuerdos de añorada juventud, los pocos supervivientes de aquella generación"* ["solidly fixing in prints which we few survivors of that generation preserve with pious care as mementos of the youth to which we long to return"] (*Recuerdos* II, ch. III). There is a telling mixture of values in these remarks: he was simultaneously a living reminder of past activities who had gone on to other activities, and the producer of images that record those times past. Cajal found himself in that cult of the Gaster Club and its dedicated members united around feasts and tables, shorelines and sailing vessels, but he also juxtaposed the present with the past as a temporal "survivor" as well as future-looking scientist.

He devotedly recorded the fields, forests, valleys, and ruins of his travels, almost always populating any landscapes with human inhabitants who, as Barthes reminds us, stand in for the now-past time of having been there to capture the image. The double images of fig. 3.2, *Cuatro Caminos*, reveal two similar shots of the tiny canals of Cuatro Caminos on the outskirts of Madrid in the early twentieth century. They appear as a backdrop in the shadows behind Cajal's posed children. In stereoscopic form, they capture the light and darkness of the land where the expanded campus of the university will soon rise. It is a place devoid of human habitation, a source of quiet natural landscape that would burgeon with life later on. Both of them—the rivulet and the children—will change through time as the city moved beyond those

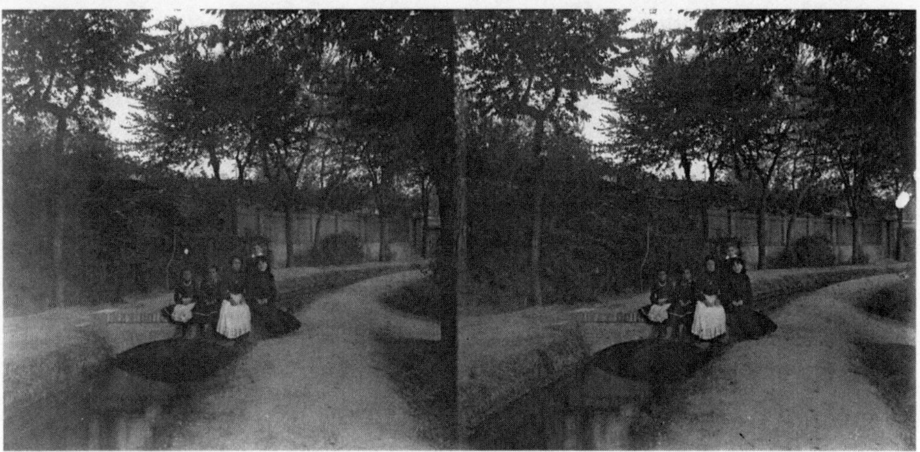

Figure 3.2. Cuatro Caminos [*Cuatro Caminos Area*]. Santiago Ramón y Cajal. Legado Cajal (CSIC). Instituto Cajal. Madrid, Spain. Used by permission.

borders, as the university area proceeded to develop, and as the modern era moved forward. The photographer and his children are captured as having "been there" at an earlier, lusher time, amid the shadows and quiet of a glade, for the eyes of inhabitants many years later.

He portrays those native to the region, those passing through, or those posed by him as part of the natural setting, the naturalness of the entire scene like a still life in which "nature seems to spontaneously produce the scene represented" (Barthes, "Rhetoric of the Image," 45). As composed as a scene of his sister and caregiver Paola in the mountains or his daughters in front of a stream might be, the Cuatro Caminos photograph also shows the number of times that Cajal recorded his family and relatives in a serious attempt to merge human beings and their habitat in unison and harmony. This vision of nature inhabited by recognizable human beings—his sisters, colleagues, children, and wife are well known to the observer from so many previous photographs—documents their having been there at some time. But it simultaneously implies change—the having been there but no longer being there—since the photos are part of an album or collection that brings all the places into a new space and into simultaneity. All times and places coexist.

Not burdened with an idle mind in any sense, and finding potential data about the world available from all methods at his disposal,

Cajal remarked that the laboratory man should preclude isolation from the world around him. He should educate himself with information from colleagues in different fields and enlighten others with advances in the scientific disciplines in language accessible to all (*Recuerdos* II, ch. IV). One of those languages was photography. Given the monumental shift accorded to the value of visual representation witnessed during his lifetime, Cajal used the image to mediate the changes he saw around him and allow objects embedded in urban and rural landscapes to speak to observers. On the cusp of the new and modern, Cajal's work both reflected an artistic tendency toward mimesis—the aura of originality and authenticity surrounding home and family, for instance, that placed Silveria in her customary chair with the children seated near her—alongside an obvious constructedness of place such as found in a still life.

Taken as a body of work, Cajal's evidence proposed the technology itself as the subject of representation. The microscope could, of course, provide a montage of fragmentary parts that, when assembled, might form or reveal previously unseen structures rather than recaptured sameness. In the camera, Cajal found similar possibilities. He became as intrigued by the production of lithographic plates as by the game of chess, an activity that "*amenazaba seriamente mis veladas*" ["seriously menaced my evenings"] (*Recuerdos* II, ch. IV) and turned into what he deemed a vice. He was as captivated by the structure of the spinal cord as by the retina of the human eye. There can only be so many hours in a day, however, so a combination of the photographs, the slides, the culinary outings, the chess games, and many other intellectual activities all portray Cajal as "having been there" in a material sense and having left just as a material record of time passing.

There occurs more of a sense of immediacy in these images, however, than with the vaster and more complex codes of linguistic representation, a fact reflected in the essays of philosopher José Ortega y Gasset, written during the same period in the early twentieth century. While Barthes concludes that the photograph "corresponds to a decisive mutation of informational economies" ("Rhetoric of the Image" 45), since it differs from both written linguistic signs and motion pictures, language connects to bodies of information and entire linguistic systems with associations embedded in those larger schemas. Those informational economies of the photograph and other technologies of modern times will cohere into the codes of new ways of seeing.

As the previous chapter established, both Cajal's histological draw-ings of the components of the human brain and nervous system and his devotion to new techniques for the reproduction of color and light in photography converted science into art, and night—obscurity, the lack of an easily visible image—into day. The emergence of an image from its surrounding field of luminescence paved the way to looking into darkness for hidden knowledge. Light itself was the sub-ject of great experimentation in many fields in the 1870s and 1880s, leading to Thomas Edison's perfected carbon filament in 1879, which ushered in the modern age of electricity.[1] Although metaphors of light have been powerful throughout Western culture—as "a directed beam, a guiding beacon in the dark, an advancing dethronement of darkness, but also a dazzling superabundance" (Levin 31)—in the case of modern technological innovations, the use of projected light for the actual physical illumination of objects under a microscope or in a captured image responded to the possibilities of a particular mod-ern historical moment. Electric light and the new architecture "*de hierro y hormigón*" ["of iron and reinforced concrete"] (Guerra de la Vega 127) were the staples of the construction of Madrid's Gran Vía, especially during the 1920s. These were only two of the many public facets of modernity that characterized the decades during and after the turn of the century. Sound cinema and radio both emerged as new technologies then also, with foreign investment in Spain coming from the United States (Western Electric) and Germany's Telefunken, which saw Spain as a potentially prosperous market. Interest in sound technologies, gramophone recordings (starting in 1888), and in larger movie theaters to accommodate the public demand for spectacles with sound accompaniment (Guerra de la Vega 175) supported construction projects that could be realized with the new architectural strategies and whose facades could be lit to attract crowds. Light propelled entertain-ment as well as scientific research.

If sonority—square, shop, church, balcony, courtyard, café, feast day, pawnshop, lotto, auction, bazaar—pervaded the city of Naples for Benjamin and his friend Asja Lacis, a different, more mechanized, sound accompanied Cajal into the landscape of *madrileño* modernity. In his travels to Columbia University in New York on scientific business, Cajal openly acknowledged his curiosity about "*las nuevas invenciones industriales del pueblo más genialmente dotado para el cultivo de la mecánica*" ["new industrial inventions of the nation endowed with the greatest genius for mechanical development"] (*Recuerdos* II, ch. XVII). He lost no time in examining the advances in Edison's work on light and on the phonograph. Writing of himself in the third person, Cajal adds a note related to his own work on technologies of reproducing sound: "quien esto escribe incubaba también, por entonces, cierto perfeccionamiento de la máquina parlante" ["the writer of these lines at that time also was hatching ideas for the improvement of the talking machine"] (*Recuerdos* II, ch. XVII). Yet he calls his unwillingness to accept radical changes in Edison's version of the phonograph a "radically egotistical" failing (500) of many inventors. He goes on to explain how the personal had an effect on the technological as he recounts—in yet another digression—the "furore in Madrid" during the 1890s over the arrival of the Edison phonograph. While Cajal lamented the unscrupulous—the term is his—criticism of artists and politicians that arose among the citizens toward those who dared to record their music and harangues, the scientist recognized that this invention had become the object of a "cult" (500). It is somewhat peculiar that the phonograph's cult status received his censure but that of the scientist did not. Perhaps the economics of each created such a division. The extraordinary value accorded new inventions, constructions, and technologies led to their elevation to a higher status. Benjamin counseled against a new auratics of items and images in place of the cults inherited from tradition that these might help sweep away. The lingering of a historical past in urban spaces mixed among the new constructions—Benjamin has in mind the Parisian arcades (Caygill 74)—fused memories and the disruption of them with the possibilities of reorganizing experience and not just a commemoration or mourning of what had passed. Cajal seemed enthralled with Edison's experiments. Despite his economic failure in the case of the gramophone, those improvements did establish the scientist's intellectual suppositions as valid theoretical bases for what others would implement. Not so the popular acclaim.

Having made his brief commentary about the cult value of modern technologies, Cajal turned to what he really wished to discuss: the deficiencies of the device's wax cylinder. Just as he had deemed some scientific conclusions insufficient or even incorrect when compared to nature, he found the timbre and volume of the voices recorded either too weak and unnatural or too strident and harsh for the human ear.

As always, he reverted to scientific investigation to correct the defect: "*Previo análisis minucioso de las condiciones físicas*" ["After a thorough analysis of the physical conditions"] (*Recuerdos* II, ch. XVII) related to this defect, he detected that a shift from recording on a groove to on a flat cylinder would lessen the noise level. An unskilled machinist turned out a badly manufactured version of his enhancement which, set aside in the attic, was forgotten while other inventors successfully marketed the gramophone that held similar changes in structure. Cajal concluded that each professional should stick to his field:

No por vanidad pueril refiero estas cosas, sino para que mis lectores biólogos, médicos o naturalistas, aprendan a mi costa a no malgastar el tiempo persiguiendo invenciones fuera del círculo de los propios dominios. Al abandonar el tajo habitual chocamos siempre con el escollo de ignorar o de conocer somera o incompletamente los antecedentes bibliográficos e industriales (patentes de invención registradas, etc.) del asunto, así como la labor intensa y sigilosa desarrollada por hábiles ingenieros a sueldo de los grandes establecimientos industriales de Europa y de América.

[It is not out of childish vanity that I recount these matters, but so that my readers who are biologists, doctors, or naturalists may learn at my expense not to squander their time in pursuit of inventions outside the sphere of their own areas. When we abandon the usual furrow, we always run aground on the rocks of knowing only superficially or incompletely the bibliographic and industrial background of the subject as well as the intense and silent labor carried out by skillful engineers in the employ of the great industrial establishments of Europe and America.] (*Recuerdos* II, ch. XVII).

Once again, Cajal found Spain lacking in financial support or skilled modern craftsmen capable of getting such inventions produced. Financial considerations were not at the top of his list of priorities.

At the end of this chapter of Cajal's recollections, he extolled instead the virtues of chance rather than the persistence of that "intense and silent labor" of the great industrial nations. The Spanish scientist, it appears to him, should be cautious:

> conviene desconfiar mucho de las invenciones de sentido común. ¡La lógica es don tan corriente, tan generosamente repartido! Y aunque sea humillante para el orgullo del investigador, fuerza es confesar que sólo los hallazgos casuales son completa y absolutamente nuestros. ¡Precisamente aquellos en que menos parte hemos tomado! . . . ¡Oh, el azar venturoso, la musa de los perseverantes y pacientes! . . . ¡Cuántos que pasan por genios te deben sus mejores conquistas y el halago embriagador de la notoriedad! . . ."

> [it is wise to greatly distrust inventions of common sense. Logic is a gift so commonplace, so generously distributed! And though it may be humiliating for the researcher, it must be confessed that only chance discoveries are completely and absolutely our own. . . . Oh, fortunate accident, the muse of the persevering and the patient! How many who pass for geniuses owe you their greatest conquests and the intoxicating flattery of fame.] (*Recuerdos* II, ch. XVII)

With a lifetime dedicated to the laboratory and the dark room almost behind him at this point, Cajal praised the personal and professional traits that had led him to cult status as a model *Spanish* scientist. He also recognized the need for refining and replacing the scratchy, ear-piercing speeches and melodies reproduced through the phonograph with more faithful reproductions of the sounds of the human voice. Truth to nature remained at the basis of his scientific work.

The same decades saw photographic processes and images reproduce detailed studies of nature on paper instead of metal or glass, with altered and enhanced effects of light and dark contrast. Casting light on microscopic structures revealed new dimensions of life as much as it required changes in theories and raised new fears and excitement about the unknown. A plate or a photo not only captured the enduring image of an object but it took that same object far afield from its original context. In 1888, Cajal not only envisioned the use of such technology in the scientific laboratory where his study of the retina (insect, animal, and human) revealed the organizational principles of the nervous system and of its optical components, but he subsequently found the effects of light on the photographic image an important corollary. The process by which animals and humans translated light into images in the brain brought him closer to refining the procedures for focusing light through lenses and on different surfaces. If he studied neuronal landscapes, replete with metaphorical references to their trees and branches, he also mapped the landscapes of Spain in his move from Zaragoza to Madrid, and he recorded his travel to the United States, Italy, and other European countries. And what could be more appropriate than a man of science using scientific methods associated with modern times to record the making of the nation and its capital, Madrid? What was occurring in Madrid belonged to the same urban modernization that included material objects such as the camera. Photography is an indicator and witness of radical change: it dissolves the notion of the perpetual and proclaims the advent of a challenge to and dissolution of what seemed solid and permanent.

Some nineteenth-century pictorialist photographers revived cultural codes that idealized a past romanticized or monumentalized in both the rural and the urban landscape of the nation. An emphasis on subjective emotion drove them to embed the human figure in the mists of time, in verdant fields that would become spaces of construction, in riverbanks that would dry up, and in the exotic architectural remnants of Spain's remote past. From his direct observations, Lee Fontanella determines that evidence shows that the convergence of the photographic medium and the notion of landscape implied a combination of two sets of conventions: "Although for many photography is still . . . a realist phenomenon, landscape is invention" (163). The lens of an instrument derived from scientific experimentation recorded a contrived image, either a constructed scene or the discovery of one that corresponded to the photographer's desires. The search for progressively more clarity of the images under the microscope, as well as of portraits and landscapes, reflected technology used at the service of conventions that the photographer-scientist had absorbed. Yet those images could be placed at the service of affect rather than knowledge, of profit if mass-produced, bringing the unknown or lost closer to the eye. The photo as a mechanically produced image was accorded emblematic power to capture people and places best through purportedly noninterventionist means, but the pictorialists found in photographic processes ways to intensify expression for optimal effect.

While photography and microscopy illuminated objects brought into their focus, the choice of angle and definition contributed to the production of the final image. Fontanella continues, concluding that the confluence of scientific innovation and a pictorialist vision led to an "enhanced *value* of landscapes" (164) created for consumers. Manipulating the image to make details less distinct, evoke the exotic (as seen through the eyes of the tourist or traveler), or create the illusion of a more documentary moment can change how the public reacted to a photograph. As Cajal traveled through Spain, he appeared to insist on the actual moment, not to capture loss but to record ongoing time (recuperation from illness, moving to the city, etc.). Benjamin referred to these images as remnants of "the stream of real life" (cited in Caygill 75). Cajal composed his encounters with places but did not revisit them as a photographic imperative. He rarely returned, but instead moved on to accumulate new images and add them to the album. Just as the landscapes were not inert but part of historical processes, the eye of the photographer and his relationship to the camera advanced with the times.

Products of observation, objects and technologies entered the modern world as so many mediated forms of the natural and the constructed (urban landscapes, architecture). As Leach summarizes in general about inventions of radical change, "[Walter] Benjamin's modern world begins with the machine" (9). It is a period that ushered in both anxieties and dilemmas that haunted civic discourse. And science, like architecture or geography, can "enter modernity as equipment" (Leach 16) as well as product. The camera was as much a part of the collection of equipment as were the steam locomotive, the telegraph line, the streetcar, or the building material of the apartment building. The relational values of an "acculturated natural setting" (Leach 18), the surfaces of new landscapes created from the world of nature and that of humans, change.

How do the observer and such products dialogue with history in what Walter Benjamin calls the passage into modernity as a "crisis of the object" (cited by Leach 6). This is a crisis of the experience of the object when mediated by new technology. The "transparency" of linear time and space are suddenly disrupted by simultaneity, montage, and Benjamin's sense of "porosity" (Caygill 39, 40), or the chance encounter with circumstances and experiences. Instead of a unity of subject and object (observer/city; eye/landscape), Benjamin described such mediated categories of modern experience with the terms *threshold* and the *shock* derived from "the impure dispersal of anonymous transitivity" (Caygill 120) as the observer made his way through the changing space of the city, which provided a bombardment of confrontations and provocations. Image and experience—past, present, and future in memory, evocation, and material contact—in unstable combinations made the apprehension of time and space not fixed but as transitory as the traveler's passage through them.

A unique individual yet part of a greater structure—a cell among others, as he has recorded in his memoirs—Cajal also became witness to the throngs of city dwellers whose urban landscape was being transformed by the construction of new buildings, parks, avenues, public works, and the simultaneous disappearance of earlier landmarks and features. His interest in the tonal values of color photography during the last decades of his life pushed Cajal's photos beyond mere contrasts between light and shade into the realm of multiple colors and contrasts. Only the complexities of the natural world challenged the power of the lens. Cajal found material in rural landscapes, as well as the streets of the city. Caygill writes of Benjamin's philosophy of the modern metropolis at the beginning of the twentieth century, "the experience of the city replaces [for him] substance and subject with *transitivity* . . . Instead of being conceived as a finite number of forms which anticipate and govern the shape of experience, the categories are now seen as intricately woven into the weft of everyday life" (120). Photographic images of travel and transit recorded the experience of time as much as they memorialized equipment.

For Benjamin, as it seems was true for both Cajal and Ortega, the paradigm of the eye's experience was more "chromatic" (Caygill xiv) than just the duality of light and dark, more a complicated mixture of color, pattern, tonality, light, shadow, and texture. This chromaticism replaced the more traditional relationships between inside and outside, space and time, continuity and disruption, opacity and transparency, dark and light, even orient and occident. Instead, "a mobile, dialogical quality" of the modern experience of landscape (Caygill 65) replaced a one-way vision in the eye of the observer. What was formerly an inert, silent scene "was brought to speech" (Caygill 65) by the intervention of the photographer and by the intervention of photographers and their cameras, travelers and their words, wanderers and their recollections. Benjamin and Lacis extend the notion of chromatics. With reference to descriptions by others that change the colors of Naples's stone houses and streets from darker tones to something more inviting, they write: "Fantastic reports by travelers have touched up the city. In reality it is gray; a gray-red or ocher, a gray-white, and entirely gray against sky and sea." Fantasy aside, although that word hints at the influence of the observer on the representation of what is seen, it is only a careful eye that makes out forms, penetrates the architectural dead-ends, or acquires knowledge from the Neapolitan. Benjamin adds: "anyone who is blind to forms sees little here" (*Reflections* 165). The vision of the observer has to learn to see what there is, to interact with the forms and shadows, the light and darkness, of the surrounding landscape. Ortega is similar to Benjamin in his philosophical interpenetration with observed landscape.

Ortega adds to Benjamin's notion of orienting one's sight to the light and the terrain of particular spaces. He leaves Madrid by train for Asturias, trading the geometría de la meseta" [geometry of the central plain] ("De Madrid a Asturias" 80) where traditional concepts of the vertical (trees) and the horizontal (distant horizon) contain the omnipresent ruins of churches, towers, and walls, for a space where it is harder to make sense of what is seen. The reds and golds of the terrain in the setting sun, "*la plenitud a que llega cada color convierte a todos los objetos—tierra, edificios, figuras—en puros espectros vibratorios, exentos de pesadumbre y corporeidad*" ("the absolute plenitude all colors reach turns objects—earth, buildings, human figures—into pure vibrating specters, emptied of weight and materiality") ("De Madrid a Asturias"84), turn into sheer absence, "*un fracaso visual*" ["a visual failure or a failure of vision"] ("De Madrid a Asturias" 85). It is as if blindness has set in. Citing Darwin's statement that in the flatness of the Argentine *pampa* the wanderer's footprints in the sand create the trail itself, Ortega finds that there is no longer a point of orientation for "*la pupila castellana*" ["the Castilian eye"] ("De Madrid a Asturias" 84, 86) on entering Asturias, but instead a turn to physical interaction. He cites a need to reach out and grasp the materiality of the objects behind the "*espléndida piel cromática extendida sobre las cosas*" ["splendid chromatic skin spread out over all things"] ("De Madrid a Asturias" 87). The eyes of Benjamin and Lacis touch the surfaces of Neapolitan architecture and geology where "the stamp of the definitive is avoided" (*Reflections* 166) and a fleeting relationship comes into being. The eyes do not provide such information for Ortega. Ortega sums up this experience as a landscape acquiring "porosidad, las piedras no acaban donde acaban, sino que en sus poros penetra el azul del cielo y el bermellón de los terrazgos" ["porosity, stones do not end where they end but instead their pores penetrate the blue of the sky and the cinnabar of the earth"] ("De Madrid a Asturias" 89). In the same way that Naples is an abstraction brought to life by the traveler's eyes, Ortega ends one portion of his essay with a plea for the need to separate the political and historical construction of "Spain" from "*una imagen visual adecuada*" ["appropriate visual image"] inhabiting the imagination of all travelers. The album of images disappears in favor of the mental image, the collection of data stored in the brain of the observer. More specifically, Ortega refers specifically to "*la retina*" ["the retina"] that perceives, records, and then stores what has been seen, which can then be revived "*cromáticamente*" [in all its color and glory] ("De Madrid a Asturias," 91) when the word *Asturias*, or *Castilla* is uttered. Ortega's lens belongs to the human eye as a camera in its own right, as well as to the capacity of the brain to store information, sensitivities, and color.

The chromatics of Ortega can be found echoed in the chromatism of Benjamin, what Caygill calls "the colour of experience" as complex visual perception. The folding together of time, space, and light in experience makes word and image reflect new ways of organizing that experience in order to make sense of the rhythms of time and the conjunction of memories. What Ortega wrote of the porosity of form and color in the landscapes of Castile and Asturias, their multiple configurations of color and shape that are transitive and shifting with the turn of the earth, the position of the sun, and the contrasts of light, dark, shadow, and tone, fits well with Benjamin's essays on color and experience. In the words of Caygill, Benjamin considered that the adult intuits the meaning of the world through "the experience of chromatic phantasy [which] exceeds the forms of spatio-temporal intuition; it is the 'medium of all transformation, not its symptom'" (83). Color is therefore a medium that replaces form, seeing the world through "brightness and transparency" (84) before material form. For Benjamin, the infinite expanse of color drives the eye to light on the linear and the limited. It opens what Ortega would call the retina to a much vaster complexity of phenomena.

With the public acknowledgment of Cajal's scientific achievements came his increased mobility, and new opportunities to observe and reflect on the transformation of the Spanish landscape. Madrid is not shown as a spectral space anchored in the past but as a live topography of change, a palette of colors and scenes experienced in motion, not immune to the passage of time. Like photography itself, the city would have to respond to new generations of inhabitants and investigators. Madrid seemed to defy repeatability and became invitingly porous. What Leach calls the "wandering eye" (20) of the observer is recorded as it turned on architectural forms of the past now inhabited by contemporary Spaniards, producing the need for critical attention to the anomalies of modernity.

As Andrew Benjamin points out, there are two senses in which one's encounter with the landscape of a city takes place: the eye and language. In an attempt to record what he refers to as "the affective city" (39), time, sound, distance, and image are inscribed on the porous spatial dimensions of cities. In his city portrait of Naples, Benjamin found that "there is no distinction between what is fixed and permanent and (its feared opposite) what is transitory—rather, everything is in a continual process of discontinuous transformation" (Caygill 121). Changing form implied changing information too. Previous maps and guides become outdated and the individual traveler-wanderer is reliant on the senses to comprehend his environment. However, Benjamin studied the collector of scenes, images, and artifacts as the possessor of fragmentary knowledge, "a motley agglomeration of random finds, of *objets trouvés*" (Gilloch, *Myth and Metropolis* 89) that were stumbled on. On the other hand, Cajal preserved a myriad of scenes, objects, and people he recorded with that previously mentioned Kodak camera. He grouped family portraits, city scenes, Gaster Club excursions, and similar events under unwritten rubrics that are part fortuitous, part habitual. Benjamin considered Naples to be a place where "time and space are not compartmentalized and allotted for the performance of specific activities . . . but frames for possibilities and potentialities" (Gilloch, *Myth and Metropolis* 28); Cajal's photographs seem to present a record of Spain as equally promising if less overwhelmingly so. He portrayed other men of science with whom he worked, perhaps in the hope of inspiring such forward-looking research in more countrymen. But that spirit, if it was visible on the streets of the city, was harder to root out. Benjamin's idea of the flaneur as a stroller of urban boulevards collecting the fragments of his encounters in a kaleidoscope of fragmentary images, his own porous recollections aroused by and arousing unexpected experiences, was mediated by metaphor and language. Madrid was recorded with the cameras Cajal used. The technology itself inspired the scientist to seek new encounters, to find new vistas, to employ new techniques. Each time he was challenged with new vistas—from the sanatorium for tuberculosis patients in the mountains in the 1870s to the summer excursions to the Levantine shore, from the provinces to the developing capital—Cajal opened the lens of his camera, striving to capture changes in himself, in his surroundings, and in technologies. If Benjamin wrote of the "language of gestures" (Andrew Benjamin 40) as embodying and producing a material presence of a "now" in the characteristic cafés of Naples, then Cajal captured the language of faces belonging to the urban streets. The physical presence of the observer offered an inexhaustible potential for storytelling to all three: Cajal, Benjamin, and Ortega.

As Barthes concluded regarding the differences between the painterly and the photographic, "Painting can feign reality without having seen it. Discourse combines signs that have referents, of course, but these referents can be and are most often 'chimeras.' Contrary to these imitations, in Photography I can never deny that *the thing has been there*. There is a superimposition here: of reality and of the past" (*Camera Lucida* 76). The era of digital photography with its challenge to the "reality" of an object having "been there" arrived long after Cajal, but Barthes's critical notion of the superimposition of past and present aligns well with Benjamin's goal of transformation rather than preservation in the energizing of the past by the present. In his 1912 manual on color photography, Cajal rejoiced in having contributed to the revolution of photographic processes and to the emergent quality of the images. On looking at them he was aware—in a very positive sense—that these would form part of a long history and not a truncated triumph. It was also obvious that he would not in his lifetime share in the inventions that would come from the laboratories of new generations of scientists. He addressed those that would follow in his footsteps (so he hoped): "A vosotros, los jóvenes, reserva el porvenir gratas sorpresas. El progreso de hoy se llama la fotografía en color; mañana se cifrará conjuntamente en la reproducción del color, el movimientoy el relieve. Entre tanto, aprovechémonos de la labor meritoria de sabios e industriales" ["The future has in store wonderful surprises for you of the younger generation. Today's progress is called color photography; tomorrow will include the reproduction of color, movement, and depth"] (*Fotografía de los colores* 19). As a scientist, Cajal did not revel in the idea of permanence but in constant progress and transformation, the expansion of knowledge, the deployment of technology at the service of human comprehension.

One of the chromatic (rather than merely graphic or linear) photographs in Cajal's collection—whether taken by him is uncertain since it has no label—was a panoramic photograph of Madrid taken from the Plaza del Callao around the time of his arrival in the city. The high perspective allows the eye to look out on a totality that meets but then escapes the domain of one's vision. It shows a cityscape comprising many neighborhoods and expanding into the countryside more each day. There is neither a totalizing quality to the image nor a tourist sense of permanence. Rather, he captures an expanding metropolis from the Palacio Real to the commercial center (the Plaza del Callao itself), extending in all its contradiction into the distance. Perched above the Gran Vía as it became the hub of the modern city, the image bears witness to what Benjamin recounted in his portrait of Naples: "the principle of transitivity as opposed to those of the 'subject' or 'substance'" (Caygill 121). That transitivity—transition from one moment to the next, from one place to another—provided Cajal with the technology of capturing the flow of time.

As new photographic technologies replaced old, photographic images would record as well as embody transitivity. His experience of modernity—that included a growing vision of the value of science—encompassed sights, sounds (Cajal's interest in the phonograph was evidence of this), demolition and construction, and what Benjamin called in modern cities the lingering of an historical past (time) which has now "become space" (Caygill 69). The streets were a theatrical stage created for the spectacle of modernity. As men strolled the boulevards in formal dress, or sold commercial wares on the sidewalk, the city surrounded them not as dead ruin but as part of the un-damming of a "stream of real life" (Benjamin, cited in Caygill 69), an interruption of artificial tranquillity. The throngs in the streets were not orderly but chaotic as modern crowds would be, noisy in the manner that Benjamin rejoiced in, with their excesses as signs of only one "of a number of possible outcomes" of social life (Caygill 75) of the historical moment. Boisterous *madrileños* filled the center, eager to buy goods, attend theatrical productions, find out what the modern cinema had to offer, or be part of the *tertulias* that went on (Ramón Gómez de la Serna was only one of many who held court with the greatest representatives from the worlds of the arts and sciences). In universities, researchers dreamed of ecstatic moments of serendipitous discovery that Cajal suggested could happen as long as scientists were at work. Urban thoroughfares could hold equal potential for new experiences. With his camera in hand, Cajal recorded both. As a distraction from the laboratory, and a compensation for hours of research, Cajal cultivated his relationship with Madrid: "he cultivado siempre en Madrid dos distracciones: los paseos al aire libre por los alrededores de la villa, y las tertulias de café" ["I have always cultivated two types of distractrion in Madrid: strolls in the open air around the city, and *tertulias*"] (*Recuerdos* II, ch. X) Those urban explorations complemented the character of the scientist, and exposed him to the colors of the *meseta* at different hours, and conditioned his mind to return to other work.

While Cajal aimed at the best resolution and the most faithful color reproduction in his photographs, inducing the observer to look more carefully, he also experimented with microphotography and stereoscopy that recorded differences in split-second intervals and double images. The image of a human face reduced in size so as to be only visible through a microscope was a challenge that combined both of his talents. No longer easily accessible to the human eye as when it was reproduced in more standard size, Cajal's self-portrait required the observer to employ a microscope to view what would elude easy sight. That intermediate step underscored the photographic image as "potential" if the right technological means were available. The existence of cells, bacteria, images, and who knew what else in that realm of mystery beyond the normal range of human sight opened a new frontier for scientists and artists alike. In 1888 Cajal completed and published a volume of 203 original illustrations based on his preparations of micrographic and microphotographic information. The title, *Histología y técnica micrográfica* ["Histology and micrographic technique"], reflects the scientist's dedication to method as well as to accuracy in the translation of information. What he called the "new truth, which I had so laboriously sought out and which had been so evasive during two years of many attempts, suddenly arose in my spirit like a revelation" (Freire and García-López 3). The relationship between nerve cells through contact—a scientific discovery—is described in terms of human experience.

Stereoscopy—the use of two lenses to mimic the dual perception of two eyes set slightly apart—was a technology that permitted scientists to get closer to what they had seen through the lens of the microscope. It also permitted panoramic visions of landscapes. It mimics what one might observe if actually present at the scene, recovering the direct contact lost through technology. Stereoscopic images reproduced the function of the brain in visual perception translated through the structure of the eye, but they most frequently did so with the use of modern machines (a conjunction of lenses). Side-by-side images taken a fraction of a second apart gave the illusion of binocular vision when viewed cross-eyed or through a stereoscope. What may have entranced Cajal about this technology was directly related to his work on the visual center of the brain and the processes by which images are translated. While the stereoscope entertained the public because it imitated nature and created the illusion of depth, the images produced made the act of looking a new experience to be worked at. In addition to the dual views, his increased use of Lippmann and other color processes to compose still life scenes, create portraits, and reproduce the colors of nature, enhanced his views of vivid settings in Asturias, Huesca, Santillana del Mar, and Segovia. Cajal's *Atlas estereoscópico del sistema nervioso* [Stereoscopic atlas of the nervous system] provided both images taken from his laboratory work and a self-portrait in the lab. His references to trees and branches link nature and the laboratory as much as the capturing of photographic images with scientific instruments does.

If Benjamin evoked an architectural porosity that invited the eye of the observer, capturing time through the fleeting relationships they forged, as a photographer Cajal was a transitional figure whose work reflected the changing experience of space and time in landscapes. For his part, José Ortega y Gasset elicited a radically different vision of Spanish landscape than had been brought to bear by the Generation of 1898. From their auratic sense of space and place, a communion of humans and landscape outside the ravages of time, and always capable of being resuscitated, landscape was "brought to speech" by Ortega. Colors were of heightened interest over the actual shapes of the terrain, and the eye had to learn to see new and radically different landscapes. The earth changed, as did the perspective of perception of the observer. He accorded what he saw a "mobile, dialogical quality" (Caygill 65) wherein manmade structures, geological formations, and technological devices were captured by a mobile eye. Architectural structures survived as ruins, vestiges of the past, surviving alongside buildings of the present and projects for the future. Although Ortega included references to past uses of factories and other constructs, he did not propose their revival in situ. Instead, he found the coexistence of new and old evidence of uneven industrialization and the tragedy of living in the past oblivious to change.

The first phase of railway construction—1848–1866—long past, Ortega would witness scientific modernization taking place under the watchful eye of foreign investment. As he traversed the roads and tracks installed to link spaces as they also joined times past and present in the form of monastery, factory, and palatial ruins that coexisted with urban development, he had privileged access to the greater picture. In a concept personified in the human eye, Ortega affirmed that it was by means of the "retina" that the traveler perceived his world and interacted with it. For Cajal, the retina was a structure to be diagrammed, explained, and perhaps duplicated by scientific instruments. For Ortega, the retina was a metaphorical reference to the organ of sight through which the world entered the human brain and evoked consciousness. The light-sensitive tissue of the eye, the retina is the porous surface through which the inside and the outside worlds meet and connect. Chemical and electrical processes pass across the membrane, bringing objects into view as they bring the objects observed, and the interpretive functions of the brain, together.

Ortega made insistent use of references to the eye and the retina as the two-way street that connected internal and external worlds. In a concise essay on technology and science, Ortega sums up the relationship between theory and materiality through the exemplary field of physics which unifies, through the scientific process of investigation and observation, "*el puro pensar* a priori *de la mecánica racional y el puro mirar las cosas con los ojos de la cara: análisis y experimento*" ["the pure *a priori* thought of rational mechanics and the pure observation of things with the eyes: analysis and experiment"] (Ortega y Gasset, "El tecnicismo modern" 94). Like Cajal, Ortega proposed the necessary combination of the two facets of scientific procedure to put abstractions to the test by observing the real.

In his essay "Viaje de España" ["A Trip Through Spain"], for instance, Ortega returned at least three times to reiterate the critical effects of light that passes through the eye into the cognitive system of the traveler. This holds true as well for "Temas del Escorial" ["Themes from the Escorial"], and "Viaje de España" in which the term *retina* appears repeatedly, both as a source of information and as a blind spot. Many of his travels took in sights of contention: the ruins of constructions, the reuse of old architecture for new purposes, the intrusion of capital in traditional cultures, the rupture of silence by the sounds of economic modernity, and the embodiment of time in new and sometimes radical spatial forms. As Cajal collected photographs of remote dwellings of monastic cults far from the bustle of the city but significantly imbued with recollections of other times, Ortega found in the same sort of hermits' sanctuaries in Córdoba sites of encounter between past and present. Calling them *"sanatorios de silencio"* ["healing clinics of silence"] ("Las ermitas de Córdoba"13) for the city ear overwhelmed with the sounds of modernity, Ortega was drawn into the spaces by observation, followed closely by the corollary absence of noise. Time was felt through space. History entered the eyes as spaces, as the sound of a bell that answers the call to prayer of a deeper-pitched *campanario* in the distance, as the gurgling water of a fountain, as a stream. All of these gave time physical, observable qualities.

Naples was a rich urban chaos for Benjamin, Madrid was a kaleidoscope of information for Cajal, and Ortega found in Spain's and other European landscapes images of dramatic contemplation. He discovered firsthand *"una afinidad entre el paisaje y el pueblo que lo habita, la relación entre lo construido y lo natural"* ["an affinity between the landscape and the people that inhabit it, the relation between the constructed and the natural"] (Paredes Martín 190). A forerunner of Manuel de Terán Alvarez, Ortega pronounces the interconnection between landscape and inhabitants that the geographer places at the center of his scientific measurements. All three imbue their portraits of modern landscape with more than what a circumstantial passenger or passerby might see. To them, all manifestations of the everyday are fundamental. Cajal looks out at street markets, over the rooftops into the hazy distance, and into the eyes of his now-departed children. Ortega hears the marking of time in the wind, church bells, train noise, water fountains, and even the creak of an old door opened by a monk in the twentieth century, whose appearance transports the observer back to the thirteenth. Like his plea for the cultures of modern times to think outside inherited concepts of idealism and forge a new way of seeing the world, Ortega employs an image to evoke more than an isolated moment. Time is a continuum into which all moments must be integrated in understanding.

The café society of modernizing Madrid was just part of *"el mercado del día"* ["today's (capitalist) marketplace"] ("Tierras de Castilla"35) for Ortega, but like stone buildings and the ironwork construction of the Atocha station, *tertulias* were also monuments to intellectual activity. The media and the café were sites of learning, of debate, and of openness to scientific ideas that might find implementation in the landscape of the city. Just as Cajal found clubs a necessary environment to keep laboratory scientists from falling into social isolation, he also felt threatened by the atmosphere of the café, since it would steal time from work even as it compensated for the intense life of a researcher. He found poetry in the recognizable hues of the urban metropolis—"*el gris, el amarillo, el pardo y el azul*" ["its gray, yellow, brown, and blue"]—and an antidote to the nostalgic yearnings of the emigrant for his home turf. He classified shortsighted colleagues as having "*sentido cromático de oruga para echar siempre de menos el verde mojado y uniforme de los países del Norte*" ["the color sense of a caterpillar to miss the damp, uniform green of the north countries"] (*Recuerdos* II, ch. X), and his sallies to explore the city are as vividly described as previous excursions in the country. Each terrain had its characteristic chromatic hues to be explored, and very material results came out of the scientific discussions that took place in cafés.

Recalling Cajal's early photographs of family and the poor, ruinous landscape of Aragón, and his will to science inhabiting every room of his father's house, Benjamin's essay on Naples opens with the evocation of a moment that draws the reader into a space. It does so through the evocation of an event that took place in the past. The personification of Naples's Catholicism in the figure of a priest accused of crimes against morality is resuscitated in Benjamin's reading of an architectural structure. It heralds Ortega's portrait of the opening of the *portón* ("great wooden door") on the landscape of Córdoba through the figure of the secluded monk—"*un cenobita con sayal de color de la tierra*" ["a hermit robed in a rustic habit the color of the earth"] ("Las ermitas de Córdoba" 13). Each writes of the present in terms of an historical schema from which nothing can be excluded. Far from mute figures that occupied moments in history remote and removed from today, both figures serve to posit an underlying narrative on which modern societies have been built. In other words, the immoral priest and the supposedly silent monk speak volumes about modernity. Aside from astonishing the traveler as something unexpected, "*sorprendente*" ["surprising"], Ortega invests that silence with sound. In an oxymoron, he brings together seemingly remote opposites: "para los que de ordinario vivimos en medio del estruendo ciudadano, un instante de silencio nos suena a algo cristalino que se rompe" ["for those who ordinarily live amidst the clamorous uproar of the city, an instant of silence sounds to us like the breaking of glass"] ("Las ermitas de Córdoba" 14). Can something be so profoundly silent as to have a sound all its own? The institutions of modern capital produce sounds—the noise of an economy and an ideology implemented over two centuries (Ortega, "De Madrid a Asturias" 93)—that mask subterranean architectures (the monastery) and fragile noises inaudible unless evoked uncannily by the traveler. The chromatics of Asturias, the detailed perception of the effects of light on land and buildings, the tentative appearance of a cloistered figure, the noisy crack of a crystal glass, all emerge only through that porosity resuscitated by the observer's encounter with a particular place. He must learn to see, as he must learn to hear.

For Ortega, the fountains of the German city of Nuremberg combine sight and sound into a porous encounter with "*esa otra Nurme-berga*" ["that other Nuremberg"] ("Las fuentecitas de Nuremberga,"21) embodied in the ruins of walls surrounded by the chimneys of modern factories. The new industrial city recalls for the philosopher his reading of a naturalist's theories on the triumph of one species over another. Presumably Darwin's from the previous century, Ortega brings to life similar human and scientific activity in the modernization of Germany. The continued existence of Nuremberg and its fountains draws together the crumbling vestiges of the city's past and the clamor of industry's noise in the present through the sights of rushing water. Between the hum of mass production and the peeking out of old coats-of-arms from the shadowy walls of courtyards, Ortega finds in the layers of the German city a relationship with the past he cannot find an equal to in Spain. The home of Albrecht Dürer, the source of the toy lead soldiers of his childhood, Nuremberg becomes a real, historical place. Nuremberg enjoyed a moment of glory that lives on in one last ruinous trace of its artistic past: the fountains. Personified as the young voices of the city in spite of being "*un cillero de la historia, un montón de años secos*" ["storage silo of history, a huge pile of dried up years"] ("Las fuentecitas de Nuremberga" 26), the water makes the city's past rematerialize. The look, sound, and movement of the water are similar to time. Time once again has been recovered through spatiality and sound wherein "El fluir interrumpido de esas fuentecitas enlaza la ciudad nueva y próspera con aquella otra callada hoy, próspera también un día. El pasado nos salva del presente creando un robusto porvenir" ["The interrupted flow of those fountains links the prosperous new city with that other one, now silent but also prosperous in its day. The past saves us from the present creating a robust future"] ("Las fuentecitas de Nuremberga" 27). Nuremberg is not the cradle of past culture but a laboratory of urban experience. It is not a defunct dream but the conjunction of many dreams and projects. The totality of human experience is sedimented in architecture, factory, home, art workshop, all merged into a porous whole of old and new, public and private, sacred and profane.

Between urban landmarks and vast landscapes, Ortega's word portraits capture impermanence as well as the need to learn how to observe them. These are spaces that mirror the passage of time, the moment of observation, the tense but revealing relationship between the surface and deeper—for Benjamin, hidden—structures. The intense outside sunlight blinds the observer as he leaves the train and enters a building; his retina is depleted, exhausted, overcome. To reestablish equilibrium and sight, one must adjust. To find the deeper—more precise, more revealing—patterns of the landscape of Asturias, the "*geometría de la meseta*" ["geometry of the central plain"] ("De Madrid a Asturias," 81) must be left behind as a concept of no use in new surroundings. A new sense of measurement and judgment has to be formed, a new way of looking, and a different color palette. Previous knowledge of the land has to be jettisoned in the case of "*la España multiforme*" ["many-sided Spain"] ("De Madrid a Asturias," 87), the "pupila castellana" [eye of Castile] (86) taught to see differently. The abstract—theoretical—notion of Spain as a nation is placed under direct observation as a collection of multiple and distinct parts (like the cells of a body).

In "Temas del Escorial," for instance, Ortega concludes that "La vida es precisamente este esencial diálogo entre el cuerpo y su contorno . . . El paisaje es aquello del mundo que existe realmente para cada individuo, . . . yo soy aquello que veo" ["Life is precisely that essential dialogue between a body and its surroundings . . . Landscape is everything in the world that exists in reality for each individual. . . . I am all that I see"] (49). The meaning of a nation, or the identity of an observer, emanates from that visual encounter with the habitus, the geography and the culture in which one resides. The tombs of the Escorial house the bodies of kings, but the human body—or consciousness—in dialogue with that architectural monument is Ortega's own. The traces of places and objects—past, present, and future embodied in the ruins before his eyes—bring to light how the slag heap that was part of nature was brought to life in a palace by a king. The social unconscious of what underpinned Spanish society is written on the walls of the monasteries, galleries, factories, and fountains of towns and cities. It is also embodied in other forms under the surface of the new, shiny, and visible: corridors, attics, back rooms, forbidding forests. Ortega's exhortation to look at ruins contains language similar to that of Benjamin's essays on Marseilles or Naples: "La investigación del hombre a través de sus cristalizaciones particulares constituye el nervio del libro de viajes como género literario" ["Delving into mankind through specific crystallized forms constitutes the central nerve of travel writing as a genre"] ("Viaje de España," 30). By means of travel writing and observation—the "*notas de andar y ver*" of Ortega and Terán; the *Denkbilder* of Benjamin—and photographic evidence, Cajal, Benjamin, and Ortega attest to the drive for a visible record of an optic of modernity. Their gaze was trained on the material vestiges of historical eras delineated in cultural constructs and in the interaction between natural and cultural landscapes.

Chapter Four

Science as a Two-Way Street

Contradictory Traces of Modernity in Dalí and Terán

In the *Arcades Project*, written between 1927 and 1940, Walter Benjamin looked at the panorama of modernity through telling urban fragments that revealed the contradictions that lay underneath the surface of the myth of progress. He found that increasing quantities of commodities froze history under the magic fascination of consumption, living under a spell of desire and then cast aside as newer, different objects took their place. As Buck-Morss argues so clearly about Benjamin's urban project, "this fetishized nature [of the commodity], too, is transitory. The other side of mass culture's hellish repetition of 'the new' is the mortification of matter that is fashionable no longer. The gods grow out of date, their idols disintegrate, their cult places—the arcades themselves—decay" (159). The embodiment of dreams whose day comes and goes, things that go in and out of fashion, turn to dust, become covered with the lint of time, cease to gleam under the electric lights of the broadened avenues and appear less desirable, all deposit a trace of their existence for the discerning and trained eye.

Political ideals and parties, the superstars of science and art, trends and styles, all accumulate in the mountain of historical events, material goods, and human ideals that Benjamin read in Paul Klee's painting *Angelus Novus* as a counternarrative to the myth of progress. Progress was not a one-way but a two-way street, a heap of goods and ideas piled up in the present limbo between past and future. Driven into the future inexorably by the powerful winds of history, the painted angel cannot take his eyes off what has already occurred, the dramatic and destructive source of the present and not the utopia of the future. Benjamin presents Klee's angel of history as "fixedly contemplating . . . His eyes are staring, his mouth is open, his wings are spread" ("Theses" 257). Rather than an articulated and clear vision of what has come before, this angel is looking with astonishment, speechless and awestruck, filled with an overwhelming fear reflected in his unblinking stare. Benjamin's focus on the angel's eyes, on the actual observation of all that has collected into the historical pile of experience, not some vague theorizing about it, marks Benjamin's position before the phantasmagoria of capital and his recognition of both its capacity to dazzle and its potential for challenge. Capital is overinvested with meaning by modernizing society, fetishizing its powerful symbolism for culture and investing time and finances in the production of goods that will be sold, then cast off as the new supersedes the old. Benjamin's angel does not spread its wings in celebration of what he sees; instead, he opens them and remains in place, is made smaller, less able to stand up to the onslaught of so much matter. Like the merchandise in store windows, scientific inventions and their practical commerce, cosmic theories, and stellar icons such as Einstein and Cajal, were put on display for the public to consume if they wished to be participants in the allure and seduction of modernity. Cajal as laboratory scientist, patriarch, experimental photographer, and Einstein as theoretical physicist whose image was marketed to the masses as part of their daily lives, are both indeed wish-images that belong to the "transitoriness" (Buck-Morss 159) of cultural objects. Each was appropriated by the state; Cajal fell into the hands of both left and right.

After the rupture with past imperial glory experienced at the end of the nineteenth century, coming closely on the heels of the 1888 Universal Exposition in Barcelona in which urban development and modernization were a central focus, the urge to keep modern wonders on display for the masses produced the Expo 1929, and a continued visual presentation of technology and science at the service of Spanish modernity. Yet this paean to national development and European innovation (twenty European nations participated) produced lasting architectural monuments as well as the demolition of vast urban areas that visibly signaled the speed with which projects could decay and objects and missions become obsolete. Benjamin writes of Haussmann's renovation and modernization program for the city of Paris, "behind the illusion of permanence that the monumental facades . . . sought to establish, the city is fragile . . . Transiency *without* progress, a relentless pursuit of 'novelty' that brings about nothing new in history" (Buck-Morss 96). The exhibition of technologies in architectural forms and in the cinema, in the equations and theories of Einstein on perception and the planets, in the implementation of railways and fast travel, rather than showing a road to the future are displays and memorials to the passage of time and the transience of ideals above all else. As he looks back with astonishment, Benjamin's angel does not observe individual fragments of historical experience but a collective, solid pile of discards, what Benjamin calls "one single catastrophe" ("Theses," 257). Among the remains and ruins of Spanish modernity, Salvador Dalí and Manuel de Terán make two different observations about the legacy and value of what has been accomplished. The artist looted the pile of debris for objects to recycle into the marketplace; the cultural geographer scoured the Spanish landscape for remnants of past social history to involve the possibilities of science in the lives of the inhabitants of the Peninsula. The oppositional stances coexisted in the same historical time frame.

The second decade of the twentieth century opened with the 1921 assassination in Madrid of Prime Minister Eduardo Dato by three Anarchists, violent disputes between unions and factory owners in Barcelona, the defeat of Spain in the Moroccan War (1921–23), the coup d'état of General Miguel Primo de Rivera in September of 1923 and his subsequent dictatorship, and the rise of a strong Social-ist opposition in response. As political forces were fragmenting into distinctly rival factions that would come to an impasse by the 1930s and lead to the Civil War (1936–39), Spanish society entertained both traditional and innovative visions of life. On the one hand, as Guerra de la Vega points out, "en los años 20 el toreo gozó de gran estima por parte de la mayoría de intelectuales. El mismo Ortega y Gasset promocionó, desde su prestigiosa posición, la publicación de 'Los toros' por el erudito José María de Cossió, [la] Biblia de la Tauromaquia" ["during the decade of the '20s bullfighting enjoyed great popularity among the majority of intellectuals. Ortega y Gasset himself, from his position of prestige, promoted the publication of 'Bullfighting' by the scholar José María de Cossío, (the) Bible of the art of the Bullfight] (Guerra de la Vega 24). The conjunction of the bullring with the promotion of Einstein's theories on light, mass, and energy allowed for popular audiences to have access, at least superficially, to the most important scientific discussions of the time. Using the notion of the relativity of the sun, the matador, and the logistics of the *corrida*, educators such as Pelayo Vizuete bridged the class (and pedagogical) divides between social classes through books and pamphlets written for the masses, addressing theoretical physics by means of something closer to home: the bullfight.

On screen in 1922, movie audiences could watch Vicente Blasco Ibáñez's *Sangre y arena* ["*Blood and Sand*"] in the latest technological art form. Starring Rudolph Valentino as Juan Gallardo, the story follows the life of a Spanish matador who escapes poverty only to be killed through his own love-struck carelessness in the same arena where he had earned his reputation. The message—simultaneously melodramatic and nationalistic—was carried through a medium of fascination for the masses. That accompanied other new inventions and innovations coveted by *madrileños*: gas-powered automobiles, suburban metro lines, and more modern dress influenced by risqué flappers and by designers such as Coco Chanel. Although seemingly unrelated, these factors made a new experience of movement and time visible in material ways: modern types of dancing became fashionable, speedier travel was facilitated around the city and the nation, and the political goings on in Europe were brought closer to home. Starting in mid-1923, Ortega y Gasset's publication of *Revista de Occidente* [*Journal of Western Culture*] opened the door to "los movimientos culturales europeos, . . . los descubrimientos científicos y los hechos sociales que empezaban a consolidarse tras la I Guerra Mundial. Ya el título denotaba el deseo de crear un instrumento de debate que mantuviese a España al nivel del pensamiento del mundo occidental" ["European cultural movements, . . . scientific discoveries, and social happenings that began to come together after the First World War. The title itself indicated the desire to create an instrument of debate that might keep Spain at the level of Western thought"] (Guerra de la Vega 136). In addition to publishing the poetry of Chilean Pablo Neruda and Spanish Federico García Lorca, for example, the journal included discussions of issues related to literary and artistic creation, and the possible social implications of scientific discoveries.

The 1907 appointment of Santiago Ramón y Cajal as the first presiding head of the Junta de Ampliación de Estudios e Investigaciones Científicas [Council for the Extension and Advancement of Scientific Study and Research] continued the championing of ideals set out by the Institución Libre de Enseñanza (begun in 1876). That had included the renovation of the educational system through a concentration of the best and brightest students in residential colleges and the awarding of grants for foreign study. Like Ortega's publication, such opportunities would convene intellectuals in sites such as the Residencia de Estudiantes and the Instituto Nacional de Ciencias Físico-Naturales in Madrid, as well as affording them time to study and work abroad, so they could become catalysts for change among artists and scientists in Spain on their return.

Salvador Dalí arrived at the Residencia in 1922, prepared to study in the Academia de Bellas Artes de San Fernando. Yet his appearance did not jibe with the goals of the institution, as Dalí himself attests: "Mi manera de vestir antieuropea les había hecho juzgarme desfavorablemente, como un residuo romántico más bien vulgar y más o menos velludo" ["My distinctly anti–Western manner of dress had made them judge me unfavorably, like a rather vulgar and quite hairy residue of romanticism"] (cited in Guerra de la Vega 144). Without a doubt endowed with enormous technical talent, Dalí nevertheless looked like the embodiment of an outlier, an outcast from modernity. Yet the style of his entrance exams showed a clear influence of the Cubists, a definitively European trend in aesthetic expression to which Dalí had been exposed. So more than merely personal recalcitrance, Dalí personified the spirit of the times: "Surrealism's new dawn; the search for new ways to look at the world was a feature of the period at the turn of the century, as was a readiness to break with all accepted customs and prejudices" (Weidemann 9). It was a distinctly European vision of the world. Benjamin saw them as adherents to too much of a dream world and not emboldened to awaken, but the Surrealists' elevation of Sigmund Freud to the status of leader of their subversion of art fit perfectly with Dalí's turn toward the irrational as a response to inherited values, morals, and what he saw as a blind reliance on reason. After all, the First World War and subsequent social violence in Europe had placed in doubt the promises of both politics and science. The use of scientific knowledge to manufacture weapons, not advance peaceful society, weighed on the minds of many such intellectuals and influenced their work.

Spain was not an outsider from the arena of social and political contention, and the very same institution that supported the latest trends in scientific research also promoted the development of Dalí's inner visions onto canvas. His four years at the Residencia exposed

Dalí to a variety of speakers that included physicist Albert Einstein, economist John Maynard Keynes, and architect Le Corbusier (Edwards 52); all brought to light ways of subverting inherited wisdom and traditional means of representation. Both these visitors and his readings of Freud's books turned Dalí's student days inward in search of "dream-inspired objects" (Weidemann 49) with which to give visual form to his obsessions. In his early autobiography, Dalí wrote that the noises and visions of war sent him deep into the unconscious rather than facing them. He found that "external danger has the virtue of provoking and enhancing the phantasms and representations of our intra-uterine memories" (*The Secret Life of Salvador Dalí* 31). His search for a time before—and a space beyond—the ravages of the twentieth century led him inward as well as back to Cataluña.

After World War II, Dalí shifted from psychology to science as a theme for his paintings, especially in images related to his own internalized fear of and simultaneous fascination with the atomic bomb. That devastating device appears integrated into the heads and torsos of human beings as well as a force capable of tearing holes and windows into solid flesh and bone, allowing the eye to penetrate the dark recesses of the human interior. Much as Benjamin's autobiographical fragments in *One-Way Street* temper his personal recollections through the exploration of geographic spaces like Naples, Berlin, Marseilles— what Susan Sontag writes of as "reminiscences of self are reminiscences of a place, and how he positions himself in it" (10)—Dalí embodied the dreams of science in the nightmares of the artist. The dreamscapes of Dalí and Benjamin are articulated around the notion of duality. For Dalí, multiple sides of objects, simultaneous visual perceptions, transparent bodies, and evaporating solids populate his canvases. For Benjamin, whether in language or photograph, "labyrinths and arcades, vistas and panoramas" (Sontag 10) are spaces opened up to revolutionary energy. What he critiqued as the phantasmagoria or frightening spectacle of consumerism whirled around modernity as one of its primary marks of identification. Commerce's dreamscape was strangely, perhaps uncannily, akin to Dalí's rotated figures, skeletal remains, and lands of unconscious desires. Benjamin found Surrealism in decay by 1929 rather than the heroic stance of an earlier time ("Surrealism" 226), but it was not something that could be ignored. Intoxication with science, or with its implementation through technology, had permeated European intellectual life, but now began to indicate that the relationship between modernity and so-called progress was one fraught with contradictions. Terán and Dalí embody two distinct ways of navigating these rough waters.

The intersections between science and popular culture explored in previous chapters reflect an overlapping constellation of shifting conceptions of the viewing subject, memory, space, history, identity, loss (the past), and landscape. Rather than the early nineteenth-century "tendency toward oppositional thinking," as noted by Eric Downing, setting up debates in terms of polar opposites or contrasts— "enlightened reason versus dark passion, truth versus error, . . . or subject versus object" (Downing 5)—the paradigm of integrating oppositional components materialized. As social and political conditions reached a critical tenor without resolution toward the 1930s, Spanish national regeneration was not the discrete problem it had been seen as at the beginning of the twentieth century but a search for a new integrated vision of the self (nation) that could face the realities of the new century. One example is Ortega's harmony of reason and vitalism that did not shun theory for practice nor practice for theory but instead brokered an overcoming of old categories. If we use the idea of an image, complementary dark and light versions of objects could be understood as obverses of one another and fundamentally and entirely interrelated, with one "an indispensable part" (Downing 5) of the other. Thus, the photographic image composed of both positive and negative components reconceived a comprehension of complex relations between opposites, whether in visual terms or social ones. This union of opposing forces created fertile ground for the notion of simultaneity on the one hand (Surrealism) and cultural mapping on the other (geographical and topographical schemas).

Much as Walter Benjamin's insistence on the imperative of an intimate and inseparable dialogue between contemporary ruins—structures produced by the destruction of former material structures and landscapes—and their previous architectural forms, so material traces of observation such as the photograph encompassed two parts of a shared and dynamic entity. In a sense, what formed the shadow of a material object—its negative version—accompanied that object and gave it depth, much as stereoscopic images could do to re-create the dimensions of the natural world. The centrality of vision had produced the desire for archives, collections, preservation of remnants of experience, cabinets filled with experimental slides and field samples, family photograph albums, and above all *knowledge*. By the twentieth century, what Daniela Bleichmar calls the drive to "satisfy [the] visual appetite" (301) in early modern times had grown into the use of material goods related to vision as evidence of perceived social and economic success or progress. The worth of a scientist such as Cajal was appraised by his skills in neuroscience but enhanced by his advancement of photography.

Inherited from the mid-nineteenth-century, photography and the media not only developed new methods and expedited ways of reproducing images, quickly leaving outmoded and outdated technologies in its wake and facilitating mass consumption, but it also got cast into the mode of the historical frame of the past as a traditional technology. The possible overdetermined valuation of early photography as a medium of objectivity and immediacy in scientific and other intellectual circles also coincided with what Benjamin saw as a corollary of historical time. The lengthy period of exposure required by early photographers to record an image, what Benjamin refers to as a procedure that "caused the subject to focus his life in the moment rather than hurrying on past it" (Downing 231), increased the perception of a connection between image and subject. Modern times venerated speed and contingency, notions that accompanied the rapid movement of travel, transit, and cognition. That aura of intimacy, of veracity, and of truthfulness gave way in the late nineteenth century, those years during the 1880s and 1890s when Cajal practiced photography and microphotography, to the snapshot. Such a rapidly produced image was related to a discrete moment and, therefore, to what would quickly change or had already changed by the time the photo was developed. The idea that the camera had altered ways of looking lingered, but new technologies transformed what those ways might be and how many consumers had access. Artist Salvador Dalí and film director Luis Buñuel experimented with the new technologies of the moving picture after the Lumière brothers' cinematograph burst onto the scene in the 1890s, shifting observation from the field to the theater and from static images to flickering ones. Their 1929 short film *Un chien andalou*, a Surrealist experiment in visually implementing Freud's theories of the subconscious and its irrational processes, took note of the reliance on the eye up to that time. Then the moving images reject and revoke vision. The early balcony sequence of shots and counter-shots between the sharpened razor and the cloud cutting across the moon, analyzed in depth by numerous critics over the past decades, announced a break with unquestioning (wide-eyed) observation. It just begged the audience to pause and reflect on *how* they looked at the world and *what* they saw.

Old photographs—in sepia or black and white—began to have the look of an archaeological discovery. Their subjects were no longer visible but were underlying the movable surface of the new. The capacity of multiple reproduction of an image—in newspapers and photographic portraits—would interrupt the enthrallment of the longer studio session of the past. Bourgeois consumers looked to multiple and diverse technologies as evidence of a continued way of life; leisure time for photography was no longer the sine qua non of the art. The transforming of images by technological intervention cut through the survival of a myth of accuracy and mimesis, values now belonging to a different relation between society and history than those accompanying the crisis of 1898 and subsequent loss of faith in the government. The notion of the photograph as a memento akin to the medals and scapularies awarded the defeated troops of the failed Moroccan campaign of 1921 (Ross 53–54) increased the reading of images of nobility, family, and tradition as "hauntingly empty . . . an abandoned world, . . . a world already long gone" (Downing 235) for many Spaniards. Photography depicted social obsolescence just as technological advances sped up and the look of the bourgeoisie shifted from stasis to motion and innovation. The practice of framing close-ups to remove subjects from their immediate surroundings was a way to disrupt the feeling of wholeness that a posed photograph taken over several minutes had projected. Immediacy framed the face of the subject—or family of subjects in the case of portraits by Cajal—floating outside a particular place and time, evoking a variety of readings. The abandoned world around the subject—all but ruined in its fixed and decaying splendor from the past—inhabited the album as a trace of something premodern. These traditions would be received by Dalí's generation as outmoded and no longer viable.

Well into the twentieth century, consumer goods and social practices in Spanish culture, even as the goal of Spain's rocky road to the future was not entirely transparent, offer evidence of a transformation in the status of the observer. Daguerre's work toward producing a vivid "clarity" of the image rather than its "multiplicity" (Bajac 17) at the birth of the medium of photography ceded to considerable advances in both areas with methods and techniques competing in the marketplace. The almost unlimited circulation of multiple images, recorded as a watershed by Benjamin in his renowned essay on mechanical reproduction, reconfigured the value placed on the original. It also displaced the connection between seeing and knowing that had shaped scientific practice. As photography evolved from a scientific invention that augmented observation to an industrialized medium of reproducing images, it also drew attention to "the complex and sometimes contradictory metaphors that were developed over the nineteenth century to make sense of the photographic medium and its unique relationship to truth, nature, and the visible world" (Keller 20). Ortega's value of retinal vision, the objective eye of science, the lens of discovery as a revealer of invisible worlds, produced discourses of fascination followed by what Barthes refers to as photography as "disorder and dilemma" (*Camera Lucida* 8) when the relationship between the observer and the object under scrutiny was broken by the technological device. No longer a still and contemplative gaze, the eye of the consumer is in motion; "there is never a pure access to a single object, vision is always multiple, adjacent to and overlapping with other objects, desires, and vectors" (Crary 20). The circulation of goods parallels the symbolic value of images and their fleeting contact with consumers as well as a voracious appetite for accumulation and innovation. Benjamin's observations on fashion, fragility, and the ephemeral capture the dead end of mass culture, but he sees potential for rupture of its mythification.

Photography as a medium of reproduction and of artistic creativity had bridged the gap between the laboratory sciences of physics and chemistry, the political scene, tourism, and the commercial drive of the general public. The goal of entrepreneurs was the constant introduction of new devices such as the stereoscope that added "the illusion of relief and depth" (Bajac 124) for the entertainment of the observer, that is to say the popularization of scientific instruments for everyday use. What became increasingly evident in the urbanization, industrialization, and civil disruptions of the twentieth century was the emergence of a critique of visual perception as relative, limited, and possibly even deeply flawed. The modern photographic utopia was increasingly open to debate. As Ortega y Gasset wrote in 1923 in *El tema de nuestro tiempo* [*The Modern Theme*], "A reality which would remain always the same when seen from different points is an absurdity" (cited in Malvasi 3). The hegemony of a singular vision reached a crisis, but instead of disappearing it became ever more complex and open to experimentation.

Albert Einstein, touted in numerous Spanish newspaper headlines as the man who had changed the most basic conceptions about the laws of the universe, was a pioneer figure at the center of the intersection between popular commercial culture and scientific culture. The new physics and the challenge to inherited notions of relationships between humans and the universe require receptivity to new ideas about nature. In a heartbeat, Spain joined other European nations in public discourse on the reinterpretation of the apparatus of investigation, the opening of space for the "scientific," and the broadening of intellectual horizons. The search for a new clarity about the unknown filled the public spaces of the modern city.

Einstein's discoveries, but also his persona, galvanized Spain around the notion of observed light as the power that could unlock the secrets of the universe. Einstein's touted visit to Madrid in 1923 was a conjunction between the need for Spain to be part of the public "veneration" of a genius (Montes-Santiago 113) and the impulse of a small but active group of intellectuals—physicians, mathematicians, philosophers, and writers—who found in the concept of relativity an idea that could do double service. It would satisfy the requirements of those proposing a reconsideration of inherited knowledge, and it would put Spain on the stage of the mass media. As Montes-Santiago describes the eager official invitation, written by Cajal himself in 1920, as the president of the Junta de Ampliación de Estudios at the request of mathematician Esteve Terradas, educated in Germany and eager to meet up with Einstein once again, "La seducción entre Einstein y España fue, pues, mutua" ["The seduction between Einstein and Spain was, then, entirely mutual"] (115). An old man by then, and a year into retirement, Cajal wrote in his memoirs only up to the year before the visit, but his mention of relativity and of *quanta* (Recuerdos II, ch. XXVII) are evidence of his acquaintance with both the famous man and his work. The two did meet briefly in Cajal's home (Montes-Santiago 115), but the only brief written record of that visit appeared in Einstein's diary, not in the media. The entire trip to Spain has been omitted from many records and is a tiny part, if mentioned at all, of Einstein's biographies. Perhaps more symbolic than substantial, the silent encounter between Einstein and Cajal—the visible hero and the invisible one—the few minutes spent in the old scientist's home created a bridge between Spain and the rest of the modern scientific world. As seemed to occur in most of Spain's attempts at modernizing society, this connection to the scientific "outside" was real and enticing, but ultimately fleeting.

The interdisciplinary nature of Albert Einstein's achievements contributed to supporting developments in multiple fields of knowledge including physics, mathematics, philosophy, "geological (planetary) catastrophism and fluvial geodynamics" (Martínez-Frías, Hochberg, and Rull 66).[1] Beyond these areas of intellectual pursuit, however, Einstein also left an indelible mark on the general public. The combination of experimental technologies, theoretical propositions, and challenges to old assumptions that corresponded to the decades of the 1920s and 1930s came to a head in 1927 with the introduction of what came to be known as Werner Heisenberg's "uncertainty principle" (Lindley 2). Not an entirely new concept, uncertainty seemed capable of overcoming the seemingly inexplicable with "the rigorous language of mathematics, . . . a system, a structure, a thorough accounting that would replace mystery and happenstance with reason and cause" (Lindley 2). Such is the language used by Cajal when he reminisces about his own laboratory experiments using the powerful lens of the microscope. The words he used to link his own aging generation to the potential for improvement by young scientists betray his nostalgia as well as his envy of what they might attain. The expectation of progress is there, believing fervently that "those who came after them would finish the job" (Lindley 2).

But starting with Heisenberg, the edifice of nineteenth-century science that had formed long-held ideals began to crack. Even with the broader scope of knowledge acquired about the physical world by the beginning of the twentieth century, and with the precise machinery that allowed for more detailed observation, the idea that the better the observation the more accurate the result did not hold for Heisenberg. His question sounds so reminiscent of Ortega and Benjamin, with a dose of Einstein's theories thrown in. Heisenberg concludes: "the act of observing changes the thing observed" (Lindley 4). Whether the contradictory observations of a variety of subjects could be reconciled, placed in dialectical relationship to one another, or left as disparate positions overturned many scientific convictions. But the shakeup of the foundations of the physical sciences did not occur in a vacuum. Such was the effect on the arts that the language of physics and the question of knowability permeated many areas of discourse. One result of the debates over a new concept of uncertainty was the appearance of accessibility to a formerly hermetic field for the general public. With corruption in the political arena, economic uncertainty, and an uneven industrialization, the structures of certainty that many in Spain had come to conceive of as knowable and meaningful gave way to an uncertainty of their own. Would science come to save the day as it had been idealized, or would Heisenberg throw a monkey wrench into the equation?

The May 29, 1919 total solar eclipse that put Albert Einstein's general theory of relativity to the test was no less of a critical event in Spanish culture than the July 18, 1860 total solar eclipse that left a lasting impression on then-schoolboy Ramón y Cajal. In fact, this event made Einstein a celebrity overnight, confirming through observation his predictions about gravity's bending of light. A natural phenomenon that supplemented previous experiments, the eclipse allowed for the observation of "the light from the stars [that] would have to pass through the sun's gravitational field on its way to Earth, yet would be visible due to the darkness of the eclipse" (Buchen 2). Dependent on actual observation, but supporting the foundation of previous conjecture, the natural event galvanized scientists and laypeople. From an obscure theoretical physicist, Einstein became front-page news. This included his appearance in Spanish newspapers, leading to Einstein's promotion as a popular figure consumed by the general public. Relativity, and the documentation and interpretation of relativity, took on peculiar guises in Spanish culture that included unexpected realms of daily life. As Parkinson writes,

> news spread in the autumn that the Eddington Crommelin expedition had recorded results squaring with Relativity's predictions, and Einstein's achievement was swiftly acknowledged by large swathes of the scientific community . . . a personality and biography surrendering themselves to a bewildering variety of 'genius' stereotypes, as well as the 'Revolution in Science' heralded by the *New York Times* headline of November, [and] Einstein achieved instant international celebrity as a kind of new Copernicus. . . . Meanwhile, Relativity became the subject of popular conversation, even as its increasing mathematisation made its finer details more obscure." (Parkinson 50)

From theory to observed phenomenon and scientific evidence, science shifted from the laboratory experiment to the visible world all around.

An atmosphere of reverence if not complete comprehension was part of every discourse and every product. Placing Madrid's exemplary institute of learning, the Residencia de Estudiantes, in the spotlight was one thing, but *El toreo científico* [*The Science of the Bullfight*) and other publications reveal the depth of penetration of "relativity" into the popular imagination. The application of Einstein's theories to that most supposedly Spanish of activities resulted in Luis Fernández Salcedo's *Tres ensayos sobre la relatividad taurina* [*Three Essays on Relativity in the Bullring*]. with the size of the bull, the ferocity of the animal, and the "relativity of being lame" ("Einstein en los toros" 1) contributing to an understanding of this sport in terms of what are obviously very relative scientific facts and theories. Using the word *relativity* appeared to link Spain with Einstein and with greater Europe, and therefore with a modern worldview. However, as Glick is quick to point out, "Cuando Einstein salió de la escena también lo hicieron las discusiones sobre la relatividad, al menos en este nivel [de las tertulias científicas]" ["With the disappearance of Einstein from the scene, discussions about relativity disappeared as well, at least on this level of the scientific tertulia"] (Glick, *Einstein in Spain* 312). The modern commodity and its embedding in public discourse—whether a product, a person, or an idea that resulted in a material product—did not enjoy a long shelf life.

The name Einstein had not gone unnoticed in Spain since the 1910 publication in Spanish of his critical 1905 articles in *Anales de la Sociedad Española de Física y Química* [*Annals of the Spanish Society of Physics and Chemistry*]. Yet the limited audience that would have access to those studies would expand within the next couple of decades, alongside what Parkinson refers to as "widespread, mostly underinformed discussions of physics, temporality, 'relativity,' and so on" (51). The language of science gained cachet if not the comprehension of the world that it might lead to. Heisenberg's essential disinterest in the material representation of objects and preference for "superseding them by spiritual mathematical forms" (Parkinson 52) probably set the stage for the reification of the new science in the image of Einstein. Einstein's treatise on Relativity was published as a scientific paper, but he also composed a popular tract titled *Relativity: The Special and General Theory*. André Bréton owned a copy of the latter. Other artists and writers who were concerned with keeping up with science in their own creative visions shared his interest in the theory. Parkinson calls Bréton's accounts of "the relative simultaneity in space-time . . . as personal and whimsical" (55), focusing on "states of mind" rather than the position of the observer in space and time. Yet this translation of Einstein's words for the masses allowed everyone to experience being part of what was being talked about in Europe. Einstein's focus on the movement of light waves, on optics and perception, and on the speed of light beams "seen from a different, moving frame of refer-ence" (Galison and Burnett 2) did not in fact lead to any notion of "relative simultaneity" as espoused in Bréton's words. Instead, Einstein problematized the position of the spectator, the size of the object perceived, the position of measurement, and the time factor relative to several points of reference. But that did not preclude educators, writers, and a host of other interpreters of the scientist's writings from publishing and espousing their version of this topic. It formed part of Ortega's "theme of our time" or "the modern theme."

At the more intellectual end of the spectrum, the first issue of Ortega's *Revista de Occidente* appeared in July 1923. In this journal he paid a great deal of attention to science, especially physics, and he published an essay "El sentido histórico de la teoría de Einstein" ["The historical meaning of Einstein's theory"] as an appendix to *El tema de nuestro tiempo* [*The Modern Theme*]. The confluences between Ortega's perspectivism and Einstein's ideas on relativity are explored here and in other venues. Glick has offered much intriguing and convincing evidence that "in a country such as Spain where the scientific establish-ment was very small and well integrated within an intellectual world which was likewise small . . . the conductivity of scientific ideas was heightened. Ideas spread rapidly and diffused easily across disciplinary boundaries" (Glick 248). Mathematicians, physicists, dramatists, writers, and artists came fell under the fascination of science as a topic. The modern world does not linger and instead moves on, so as rapidly as Einstein appeared in the press he ceased being the center of attention for the masses.

Einstein's trip to Spain in 1923 was a social as well as scien-tific event, one that celebrated Spain's science in the brief meeting between Einstein and Cajal who lay gravely ill. By 1930, daily peri-odicals such as *ABC* had established columns dedicated to scientific experiments and innovations, and reporters were hired to summarize the most urgent scientific news and discoveries. In December 1930, the 48-page special issue of *ABC* gave a complete summary of the year's most earth-shattering occurrences in politics, the arts, sports, bullfighting, and the cinema. Experts such as Dr. Gregorio Marañón offered an overview of medicine; Pelayo Vizuete wrote on "La ciencia en este año" ["What Happened in Science This Year"] in articles for newspapers. Vizuete contributed a great deal to the popularizing of science and to a personal interpretation of Einstein's theories. While he does not share Bréton's "disorientation" of the spectator as a parallel to Einstein's description of "separate realities experienced by observers moving relative to each other" (Parkinson 56), seeking instead some marvelously reconciled coincidence, Vizuete's *Einstein y el misterio de los mundos* [*Einstein and the Mystery of the Planets*] (1923) bridges the space between physics and the lay public to create a larger pool of citizens acquainted with the language of Relativity. That linguistic recognition does not, however, indicate comprehension of the scientific concepts used but the ability to experience the excitement of science in some general way. This meant, of course, that there were various dimensions to such discussions, from the mundane and often superficial or inaccurate to the precise, the mathematical, the intellectually founded.

The cover of this smallish paperback (fig. 4.1)—the first of several related to the same theme—with a fairly low price tag of 250 pesetas displays the nude torso of a man facing away from the viewer into a divided celestial space. On one side, rays of light emanate from the sun, and on the other, lower half, there is a densely starry sky with tiny moons, a comet, and several planets. The rings of Saturn are visible in the dark distance, and sparkling spots of light cover the deep blue firmament. The male figure holds aloft the earth in both hands, a globe that exhibits daytime on the side facing the sun and, appropriately, nighttime on the hemisphere facing the reader. Similarly, the sky holds the light of the sun and the stellar landscape of the heavens in the cover's 4" by 6" image. We are privileged to have visible access to both night and day simultaneously, revealing the power given to us, perhaps, by reading the material contained within this volume's pages. The book presumably belongs to a series with the title Ciencia y belleza para todos [Science and Beauty for Everyone] framed by mirror images of a ringed planet. A confusion begins when the heading at the top of the illustration announces boldly: "La novela prodigiosa: ¿Cómo se rasga el velo de las maravillas naturales?" ["The marvelous

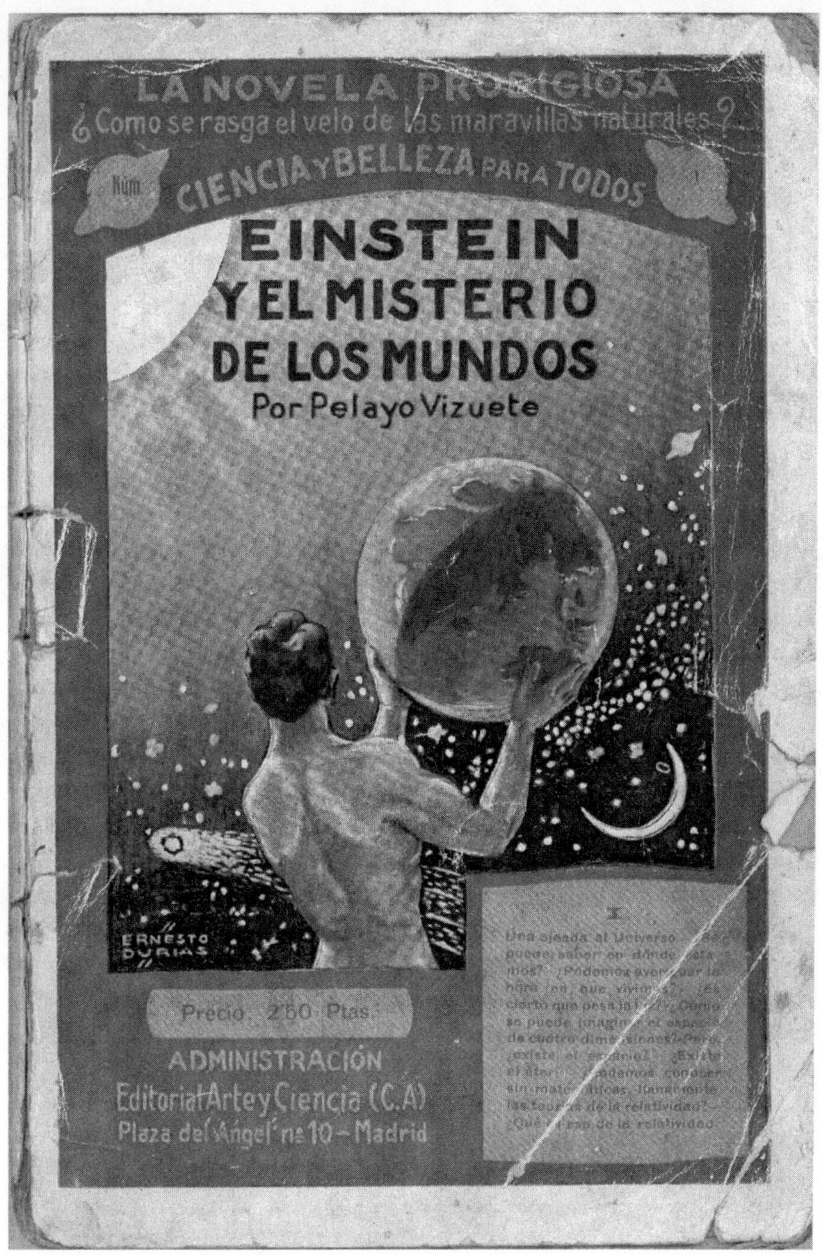

Figure 4.1. Cover of Pelayo Vizuete's *Einstein y el misterio de los mundos* [*Einstein and the Mystery of the Planets*].

novel: how does the veil of natural wonders get parted?"]. Not a work of fiction at all but proposing to explain the laws of relativity to the common reader, this tome is posited on analogies between everyday experience and the laws of science that govern our universe. While hardly fictional, the manner of narrating is more down to earth— "imagine you are standing in the center of the Puerta del Sol"—than theoretical and remote. It reflects the audience it intends to address.

Yet the combination of fiction (novel) and scientific universe (Einstein, the planets, a human holding the globe of the earth) does not occur. The catchy invitation to seek marvels, mysteries, the open skies, and some notion of fantasy connected to a novel (with a nude man thrown in for good measure) is contradicted at once by the graphs, charts, and discussion through visual analogy found between the covers of the book. Using an orange pierced by a straight pin as its axis, Vizuete studies planetary motion, rotation around the sun, and, ultimately, an imagined interplanetary voyage. With the artist Ernesto Durías as his visual interpreter, Vizuete sets out to educate the public about what the scientific community has been discussing. It is uncertain how much he himself comprehends about Einstein. Taking the Puerta del Sol as a focal point, Vizuete launches readers on a journey among the stars after abandoning their habitual home—planet earth— and watching the geographical center of Spain disappear below them. The creeping in of scientific references to space-time is reminiscent of Bréton's emotive spark that sets fire to a poetics of the marvelous. Perhaps there is more fiction than first appeared.

For Vizuete, two communities of readers coexist in Spain: scientific initiates and the average person. He sees his role as mediator, as he spells out in the "Advertencia" or word of caution and advice to readers.[2] His introduction to the series marks these divisions clearly and indicates how Vizuete plans to proceed.

> En estos libritos no intento decir novedad alguna a las personas instruidas en las ciencias relacionadas con los descubrimientos de Einstein; a tales personas las supongo enteradas, por deberes de profesión o por mera curiosidad científica, de las notables conclusiones a que ha llegado el célebre físico alemán, aparte de que tal intención se hallaría muy distante de la cortedad de mis fuerzas. El propósito que me estimula es más modesto: se reduce a explicar con

sencillez las teorías denominadas de la relatividad, para que
toda persona de regular instrucción y aun de mediano
entendimiento pueda comprenderlas y explicarlas a su vez
en nuestro idioma corriente. Acaso algunos lectores tengan
por exageradas la llaneza del lenguaje y la vulgaridad de
los ejemplos. Para ellos repito la advertencia: pueden apar-
tar de sí el librejo, que se ha escrito con el fin de dar una
explicación a los modestos de inteligencia y a los curiosos
de saber que no tengan noticia de semejantes cuestiones o
la hayan adquirido muy confusa. . . . (5)

[In these little books I have no intention of saying anything
new to those of you already educated in the sciences related
to Einstein's discoveries; I will take for granted that such
persons will have already heard of the noteworthy conclusions reached by the celebrated German physicist, whether
they have done so for professional reasons or mere scientific
curiosity. In any case, my own abilities could not reach the
level of such a readership. The goal that induces me to this
task is more modest: it is to explain as simply as possible
the theories related to "relativity" so that anyone with an
average education and even with average intelligence might
comprehend them and then be able to explain them in
their own words in everyday Spanish. Some readers might
find the simplicity of language and vulgarity of examples
too exaggerated. To them I repeat my advice: set the little
book aside; it is a modest attempt at explanation not meant
for you but for the average person curious to know about
these kinds of questions but who haven't had the chance to
find out or those whose knowledge is unclear. . . .]

There is no doubt that the author considers his task one of mediation
between a scientific community and a community of readers for whom
visual diagrams, allegorical explanations, and a good deal of repetition
are central to getting ideas across. Indeed, they may miss the mark
anyway. When addressing speed (velocity) and position, he turns to
short phrases. These include: "Empieza el misterio: ¿en dónde estamos?"
[The mystery begins: just where are we?"] (29) or "En movimiento"
["In motion"] (67) to adapt difficult concepts to an audience. He also

uses the collective "we" form of the verb to make his own voice part of a dialogue with the reader. Vizuete offers the example of German pedagogues who have successfully taken Einstein's theories into the classroom and asks Spanish readers why they should not be capable of the same. Although Cajal did his own mediation of scientific work on photography to educate those nonprofessionals who were entering the field, Vizuete took on the task of interpreting Einstein for the general population. Both Cajal and Vizuete saw Spanish society as "catching up" with Europe but still in need of their assistance to become equals. Vizuete writes: "no nos hallamos ya tan lejos de la ciencia y de la actividad pedagógica de otras afortunadas naciones" ["we no longer find ourselves so far from the scientific and pedagogical education of other fortunate nations"] (6). Yet if he was initially successful, it was a conversation that the general public soon forgot even after seeing the alluring "veil over natural wonders" parted on the cover of a popular paperback.

During the 1920s and 1930s in Spain, visual evidence to prove modernity was everywhere. As Mendelson summarizes, "documentary images, especially in the mass media, became a symbol in Spain of modernity, . . . [connecting] their depiction of local issues to international trends in the visual arts . . . and at the same time synonymous with tradition, time capsules of history" (xxi). Mendelson goes on to link such images with debates over national identity and patrimony, offering the committed spectator evidence of social practices and political issues emanating from specific regions or from Spain linked to the larger European context. But the authenticity of those images, their truth-value and accuracy—as Benjamin had pointed out already in his writings on the aura and the auratic after mass production—was also turned over to the viewer with potential power. The "art of intuition, a glance" (Mendelson xxii) that opened the domain of images to the ordinary spectator produced the possibilities of multiple readings, collective connections, and the projection of fears and anxieties onto the same visual images.

Accessibility to discourses on relativity, and to newspaper photos of Einstein among scientific *madrileños*, brought home the world of current events through the use of allegorical allusions to the bullfight and to recognizable geographical spaces of urban Madrid where one might catch the scientific fever of the age. Yet Cajal had also found in the photograph of his own documented service in the failed Cuban expedition an image that led in other directions and to other interpretations. The photograph of the young doctor in uniform did not provide him a direct view of what had occurred but one that captured his innocent gaze on departure, a gaze that now looked both disconnected and naive. His uniform documents the impulse to service, yet the result of that idealistic (and self-serving to get him out of a claustrophobic town) trip opened his eyes to the realities of both uncontrolled disease and failed colonial ventures. A single image was made relative by the memory of experience as well as by the elapsed time since its production. Cajal would suffer the physical, intestinal results of his misreading of a situation for many years. In fact, it appears that his visible weakness during Einstein's brief visit is a direct result of those lingering maladies that returned with a vengeance in old age.

Dissidence could be fabricated through the widespread dissemination of mass manufactured images. Technologies of the eye and fixed or mobile images of identity were entwined: "By the 1930s, what had begun as an experimental form of representation practiced by the few was now in the foreground of everyday life" (Mendelson xxv). This phrase sums up much of our discussion to this point. Scientific experimentation had relied on the supposed accuracy of observation and representation. But the expansion of photographic techniques and experimental methods had progressed, and the image became ambiguous, political. Even scientific images were filled with contention as to dye, contrast, light, and other conditions of the laboratory. Observation was a power that could be harnessed and deployed for different uses.

On one side of the coin, the Surrealists in the 1920s and 1930s took up the cause of attenuating the reliance on vision that had held the culture of the West in thrall for so long. Their evidence came from firsthand experience with images of war in the European theater, a war that they held to be a direct outcome of scientific inventions and technological advances used to end life rather than improve it. On the opposite side of the coin emerged Manuel de Terán Alvarez's renewed confidence in the personal experience of the geographer "de tacón gastado" ["with worn-out heels"] (Anes 10). His science of walking and observing emerged from the writings of Ortega y Gasset and the detailed study of "acculturation and landscape humanization" (García-Ballesteros 10) rather than the classroom or the laboratory. Knowledge could begin in the archive but the "science of landscapes" (García-Ballesteros 11) required the naturalist to be a humanist in contact with the objects of study. Human subjects whose constructs and choices made up the material foundations of urbanism had to be observed in place.

One aspect of the crisis in perception—not of the technologies associated with the making of images but with the interpretation of the products gleaned from it—evolved from an avant-garde questioning of mass production and potential falsification. When asked by Otto Hahn about his interest in Surrealism, Marcel Duchamp responded with a comment that brought the processes of perception and interpretation into line with debates over accuracy, authenticity, and truthfulness. No longer was there an automatic truth claim associated with the visual. Duchamp's answer in their interview reveals an important link with the eye that had been at the center of the representation of the real for so long.

> —Es el único movimiento del siglo. El único que haya pretendido dar un aspecto técnico a la pintura, que haya querido quitarse de encima su aspecto visual—no digo retiniano, porque hablo demasiado de este concepto, en cada entrevista hablo de mi rechazo a la pintura retiniana que sólo se dirige al ojo . . . Los surrealistas buscaban librarse de este aspecto sensual y superficial. . . . (Hahn 3)

> [—It's the movement of the century. The only one that proposed to give painting a technical aspect, that wanted to get out from under its visual aspect—I'm not going to say its "retinian" qualities because I have spoken too much about that, in every interview I speak of my rejection of retinian painting that directs itself only to the eye . . . The Surrealists sought to free themselves from this sensual and superficial aspect. . . .]

Addressing something not *in place of* the eye but reached *through* the eye across the light-sensitive inner surface and the optic nerve to the brain sends images, like some cultural neuronal code, to be decoded and not merely absorbed as is. Dalí's eye was unreliable as far as offering a rational vision of the world. But it was the only sense that gained access to the inner world of the brain, and that was done through irrational forms.

Cajal carefully charted the biological neuronal process, but photographers and artists described the necessity of using the same method for the reception of images in new circumstances. In an allegorical sense, Salvador Dalí's personal album of Spanish modernity included a systematic collecting of visual ephemera with which he "interrogated the institutional and individual motivations" (Mendelson 188) behind representation to evaporate any notion of an innocent eye. The slit eyeball that opens the 1929 Surrealist film *Un chien andalou*, codirected by Luis Buñuel and Salvador Dalí, represents both a fear of blindness and a conscious image of the fascination of looking. He was acutely aware of the ideological uses of vision, and the cinema could be an even medium for the metamorphosis of objects than the canvas. Through cinematic techniques, Dalí proposed to break narrative and visual stereotypes. The expectation of a scientist to duplicate experiments and so confirm his findings were replaced by Dalí's conversion of the real into the surreal.

His inclusion of photographs, "postcards, landscapes, actions, and objects" (Mendelson 188) in a paranoid-critical deconstruction of Millet's *The Angelus*, for instance, showed how documents could be used to forge ideological interpretations of historical moments and figures if the ruins of their historical relationships are not unearthed. Always dissatisfied with the visible, Dalí would probe previous layers of paint now invisible to the naked eye in order to resuscitate the vestiges of the past buried under the surface. Dalí's excavation of the hidden figures under the surface of Millet's canvas parallels his diving into the subterranean world of his unconscious to liberate the images hidden there. He was not satisfied with accepted artistic ways of representing emotions, but was driven instead to induce the spectator to react to new images in unexpected contexts. The unreliability of the eye drove him to coax out "secrets hidden behind visual phenomena" (Weidemann 30). The combination of "critical" from scientific discourse with "paranoia" from Freudian psychoanalysis produced a method that seemed to combine the two into a modern work of art.

Ortega's diagnosis of an archaeology of the image refers to it as

Un nuevo desplazamiento del punto de vista [que] sólo era posible si, saltando detrás de la retina—sutil frontera entre lo externo y lo interno—invertía por completo la pintura su función y, en vez de meternos dentro de lo que está fuera, se esforzaba por volcar sobre el lienzo lo que está dentro: los objetos ideales inventados.

["A new shift that displaces point of view, a shift that was only conceivable if, jumping behind the retina—that subtle border between the exterior and the interior—it completely inverted the function of painting and, instead of bringing inside of us what is outside, it strove to pour onto the canvas what is inside: ideal invented objects."] (Ortega, "Sobre el punto de vista en las artes" 1)

This "resuscitation of a cadaver" (Ortega, "Sobre el punto de vista en las artes" 2) executed on the perception of the images brings objects to life over and over in the mind of multiple spectators. The images of his aging spouse and deceased son could spring back to youth and life as Cajal gazed at their photographs, even as he was aware of the effects of time on the real person. The sense of an afterimage—a sensation of something visualized even after the removal of the external referent—lingered in the scientist as an experience of time recovered by means of the photograph. Yet his work on the human body made him acutely aware that the "truth" of such perception was personal and subjective, not scientific and empirically sustainable. For Dalí, time measured internally did not have such restrictions on "the mind's images" (Weidemann 30), since they inhabited the recesses of the creators unconscious and not the outside world. They were not linked to the realm of logical perception or to any outside stimuli. Therefore, the point of view had to be only that of the artist himself, no matter how many "sides" were made visible.

Through the use of anamorphic images, encrypted in photographs or paintings and visible only from different perspectives, Dalí challenged the spectator's possibilities of interpretation. The signal of an end to a singular coherent universe of observation—or observers—anamorphosis takes the illusion of the stereoscope (binocular vision) into other fields of perception. The anamorphic image requires the organ of vision but uses that to move in other aesthetic and political directions. As the optic nerve serves to thread images crossing into the brain in order to create visual perception, so Dalí's manipulations challenge how to see, and how to read, what the observer thinks is there. No longer a recreation of the three dimensions of the real, anamorphic images confront the foundations of perception and provoke the privileged domain of singular sight. More contradictory evidence is available when an image is studied from different angles and points of view. Optical intermediaries such as the lens, and other scientific devices of the laboratory, have stepped between observer and object to produce ghosts, shadows, and duplicitous images. Unlike the complementarity of the photograph and its negative, anamorphic visions begs a lack of similarity rather than an inversion of the recognizable. Cajal would have sought to solve the mysteries of images under the lens of the microscope; Dalí sought to make them more perplexing.

Dalí gathered a few around him in the Residencia, as few as the scientific followers of Einstein in Spain. If the scientist's theories were exciting and valid *because* of their limited audience, the artist fostered a similar cult. The comprehension of men of science around Einstein was made visible by their own visibility in photos. Images of gatherings, receptions, and posed portraits with visiting scientists encircled Einstein's person as a metonymy. The source of scientific wisdom stood in for the theories; the man was more comprehensible than his science. He was more approachable as a public figure than the radical proposals he had made regarding the universe. The prestige of Einstein's persona resonated in Dalí's universe with "the conviction that the description of nature offered by relativity, confounding common sense predictions of the constitution of time, space and matter, could be mobilized in support of [his] project to engender systematic confusion in the 'objective' world" (Parkinson 191). What better for the idiosyncratic Dalí than the paradigm-changing proposals of Einstein? Science gave Dalí a method—"paranoia"—of creation that defied common sense.

So Dalí would employ the reception of radically new physics as the language of his own aesthetics and paranoiac-critical method touted as a provocation to any precepts of reason or rationality. The same old language of art—and of representation in general, including that "retinal" limitation ranted against by Duchamp—could not linger on in the transformation of observation. "Altering what it seeks to examine" (Parkinson 103), and not merely reproducing it faithfully, the imagination exerts influence on perception and intervenes in what is already a mediation, a manipulation by the photographer or artist. Dalí's engagement with relativity and psychoanalysis focused on "the contraction and extension of bodies in motion, and the idea that time passes at quite different rates in separate 'systems,' 'bodies,' or 'frames' of reference" (Parkinson 177). His "nuclear mysticism," soft watches, and use of collage technique exploded notions of unity into component parts, fragments, and the liberation of observation. Madrid's Residencia de Estudiantes and, in particular, Ortega's *Revista de Occidente* exposed Dalí and his cohorts to readings of Freud, Einstein, and Heisenberg, and the artist could graft the phenomena of the transformation of physical bodies observed through theories on relativity onto the realm of the subconscious, then onto the material products of art.

Yet the noted disjunction between Bréton's proposed "relativity" of the mind and Einstein's theories occurs once again in the case of Dalí. While the scientist employed close observation of light signals, trains, and clocks to investigate the relativity of simultaneity— "Simultaneity is relative to a frame of reference, it is *not* absolute" (Galison and Burnett 3)—Dalí turned to the flood of perceptions of his senses and his mind as unending, synchronous images. The artist's use of "simultaneity" evinced the three dimensions of space (length, width, depth); he then added a fourth dimension: time. This took the form, for instance, of "Corpus Hypercubus" (1955), wherein the unfolding of a hypercube (tesseract) is used to depict Jesus on the cross. The geometric symbol for a divinity that is not accessible to the human mind unites mathematics and religion. Floating above a chessboard, with a robed Gala, Dalí's wife and muse, standing at his feet, Jesus is both human figure and scientific enigma.

In *Slave Market with the Disappearing Bust of Voltaire* (1940), an anamorphic vision is at work in the distinguishing of, or "disappearance" of, the philosopher's bust plunked down in the midst of a slave market. A seminude female onlooker appears with her back turned to us, peering at either one vision or the other. It is impossible to discern what she sees. Does the spectator have access to the double image, or is it hidden from us? Dalí proposed such images as the making of the normal from the abnormal (and vice versa) as well as a philosophical interpretation of the consciousness of actual versus perceived reality. If Einstein and others rejected subjectivity in science, Dalí became enraptured with their notion of relativity in his own interpretation. This included Dalí's exploration of unconscious links between the observer and the observed, the human body as matter *and* energy, the relationship between awakening and the somatic, and the human creation of the scientific concept of time. Figures floating in space like Gala in paintings such as *The Madonna of Port Lligat,* or the images of melting clocks in *The Persistence of Memory*, followed by *The Disintegration of the Persistence of Memory* illustrate Dalí's emphasis on the problematics of integrating image and observer, practice and scientific theory. The double images of the early 1930s morphed into the ambiguous, multiple images of "*Afueras de la ciudad paranóico-crítica: tarde a la orilla de la historia europea*" ["*Outskirts of the Paranoid-Critical City: Afternoon on the Edge of European History*"] (1936) in which a mirror, a hole in the wall, and the open sky all merge with doorways and pathways to nowhere but dreamscapes. Liquids, solids, and vapor combine in the same space, and appear as portals to other dimensions leading to who knows where. *El gran paranoico* [*The Great Paranoid One*] (1936) sums up the iteration of Dalí's fears—the emergent figures that form the skull as well as stream forth from it—as he has alluded to so often in other paintings. These landscapes cross the borders between waking and sleeping, between observable life and a delirious observer of fantasies. Both materialize in Dalí's vision of art as an act of subjective expression against the systematization of observation and explanation.

Change, elasticity, unpredictability, the fusion of past and future—in space as well as in time—united in hallucinatory, decompartmentalized realities for Dalí. The notion of an afterimage burned into the experience of the observer repeated the artist's own experience that he had attempted to transmit. Dalí's romance with the concept of metamorphosis—emergence and submergence, fluid union, timelessness and discontinuity, composition and decomposition, transience—did not rely on empiricism or truth-value but rather on enigma, irrationality, the obscure, and advocating the relative perception of the world through the lens of madness. His famous statement that the only difference between Dalí and a madman is that Dalí is not mad epitomizes the artist's requirement for using a nonscientific lens to view the objects of the material world. That lens originates in the unconscious and it gives a whole new meaning to observation. The eighteenth- and nineteenth-century precept that "observation and experiment were . . . to work hand in hand: observation suggested conjectures that could be tested by experiment, which in turn gave rise to new observations, in an endless cycle of curiosity" (Daston and Lunbeck 3), later shifted to an unlinking of the two. Dalí's complete disengagement of experience (experiment) from recording (representation) asked the observer to replicate his new vision from which the rational had been effaced. Endlessly curious, Dalí's experiments with art did not test hypotheses but rather delved into the irrational as an end in itself. The delusions and projections of internal conflict ascribed to the paranoid observer gave new life to this type of experience. The alternate (paranoid) lens through which one represented objects came from the mind and not an external lens or filter. It did not matter that the observer share it, but that he or she experience it.

The language of science gave Dalí a way to refer to the iconography of the interior world of human beings. He could convert theoretical concepts into visual images, and mine topics as well as techniques he could deploy. Swimming between "the cold water of art and the warm water of science" (Ruiz 1), Dalí navigated his own anxieties and obsessions by not making peace with either. The same way that Heisenberg questioned the reliance on facts gleaned from the observed world as "not the simple, hard things they were supposed to be" (Lindley 4), so Dalí interrogated how much one could comprehend about things from immediate observation. The eye was fallible. Dalí's 1931 *The Persistence of Memory*, the 1954 *Disintegration of the Persistence of Memory*, as well as the stereoscopic *Harmony of the Spheres* and *In Search of the Fourth Dimension* (1979) attest to Dalí's discovery of science as theme as much as it was technique. His interest in the Fresnel lens, the projection of light, and the potential value of holography attest to Dalí's fascination with sight and its manipulation. The vaporizing of material objects by the atomic bomb gave him a way to represent the fragmentation of thoughts and dream images that populated his own subconscious. The use of Ben Day dots decomposed solid images into particles that, when perceived from a distance, suggested perforated solidity and dimensionality. Following the idea that "the scientific synthesis commonly called Unity was the scientific analysis commonly called Multiplicity" (Lindley 30), Dalí represented many Galas, in multiple forms, and many dimensions of Gala. Like X-rays, like photographic plates, Dalí perforated surfaces and allowed for autopsy-like visions of the invisible and destabilized elements into other elements as well as into far-flung components.

Dalí's artistic production during his last decades was informed by his own readings of scientific criteria. Personal perception rather than the full comprehension of facts—and if he did indeed follow the theories to their conclusions he seems to have chosen the paranoiac-critical method over the experimental-observational one—gave Dalí fodder for producing both a cult persona and an oeuvre that responded to artistic and scientific discourses. What crossed the retina was not enough, as Duchamp had laid it out, and the investigations of the new physics offered Dalí the chance at a subjective engagement with creation and discovery on his own terms. His making sense of scientific and mathematical theories was to employ the current terminology—relativity, simultaneity, atomic, metamorphosis, disintegration, *Future Martyr of Supersonic Waves* (1949), *Microphysical Phosphenes* (1950)—to enfold multiple, stereoscopic, fragmented, and often inaccessible images of personal mental states such as dreams. Rather than depicting personal experience in recognizably scientific form, or in categories advanced as clinical formulas, although such forms may be evoked, Dalí adopts the language of science to underscore experiential uniqueness. Comprehension is not his goal, neither is the possibility of repetition as proof. Exposing spectators to the dimension of the imagination is. This sets up scientific truthfulness as belonging to his subjective perception within the Dalinian universe. They "belong" because they fit the source (his imagination, his dream world) and the medium of representation (the artist). The images are "authentic" in his apprehension of them, not in the sense of providing knowledge about them or about the world.

Astrid Ruffa points out that "Dalí emphasised the experimental value of his surrealist activities . . . supporting his theory with a quotation by Erwin Schrödinger [to propose] that the paranoiac mechanism characteristic of his own method underlay the determination of the experimental choice leading to scientific investigation. The obsessive paranoiac idea, Dalí claimed, occurred in an abrupt manner and focused attention on certain objects to the detriment of others" (1). The eliding of the moment of choice in art with the moment of selection in science brings the eye of the scientist in line with the eye of the artist (or photographer) in the perception of object cognition, focus, and vision. Subjectivity and objectivity merge and spill into one another, allowing for an exploration of cinematic experimentation with the representation of overlaps in space and time ["The paranoiac-critical method actually combines the speculative plane, which claims to be objective ('critical'), and the irrational plane with its subjective nature ('paranoiac')]" (Ruffa 6–7). Experimental choice and subsequent analysis expose the objective world to the subjective states of the observer. The curiosity of speculation (experiment) and observation produced art.

But Dalí was not the only adherent to the legacy of careful observation. As Dalí was mapping the features of his hallucinatory states with rigorous and obsessive repetition, Spanish cultural geographer Manuel de Terán Alvarez was charting human geographies just as avidly, if not with the same motivation. Terán conceived a written language of experience and visual record that gave authority to his investigations, uniting feeling and seeing through inexhaustible contact with the landscapes of the city and the topographic features of the countryside. Both before and after the civil war, and inestimably interrupted by that traumatic event,[3] Terán collected, cataloged, and interpreted the foundations of peninsular culture in order to document peculiarities and readings of visible clues to different stages of development. If Madrid was a modern urban experiment, then the Iberian Peninsula was the laboratory that housed it. With a humanist's metaphorical turn of phrase[4] in the collection of studies related to "Baja Andalucía" ["the Andalusian lowlands"] published in Ortega's *Revista de Occidente* in 1936, Terán writes "Como una fina dactilografía impresa en el suelo, el plano de una ciudad ofrece la posibilidad de una identificación clara y precisa de su personalidad histórica y geográfica, de su peculiaridad más íntima y diferencial" ["Like a fine detailed fingerprint pressed into the ground, the plan of a city offers the possibility of clearly and precisely identifying its most intimate and distinctive peculiarities"] (cited in Gómez Mendoza 18). Photographs, census figures, architectural plans, historical documents, and maps joined personal walks to seek ways in which history, layers of cultural contacts, topographical characteristics, and human inspiration in the face of geographical challenges all played central roles in the life of the nation. Terán's exhaustive examinations of just how "ciclos de erosion geológica y de actuación histórica" ["cycles of geological erosion and historical activity"] (cited in Gómez Mendoza 22) fostered the Toledo of his day offer one example. He thrived on the evolution of contradictory identities that emanated from historical, economic, and political events: from the nascent tourist industry's effects on Toledo to the risks of "provincializing" versus "suburbanizing" in the shadow of a rapidly growing Madrid, change could be traced and understood in both humanistic and scientific terms.

An assiduous student of archival information in the Biblioteca Nacional [National Library], Manuel de Terán, like Cajal, spent his early years in the provinces gathering data land collecting information he would later use as a basis to plot urban distances. He traced the spokes of a wheel radiating from the old central city, whose axes were centered on economic development through the "absorption" ("Prólogo" 11) and transformation of small towns. The interplay between conserving previous structures and transforming them into suburban enclaves on Madrid's periphery was a give and take of both topographies and cultures. Terán's insistent use of the word *proceso* indicates a critical aspect of the dialogue he posited between the "*circunscripciones administrativas*" ["administrative boundaries"] ("Prólogo" 11) and his findings about the time/distance relationships measured as the decreasing domination of the city over the landscape as one moved farther from the center. Metropolitan areas did not arise overnight but developed across time. Terán charted three phases of "invasion" into rural lands with social, psychological, and cultural repercussions as much as economic ones. He offers evidence of a definitive shift in mobility from outskirts to center, with 1970 comprising the last phase of urbanization he witnessed. With those observations, Terán sat at the forefront of a shift in the field of geography from description to scientific observation. The tourist versions of Spain composed and sold as guides for travelers and the curious were replaced by more normative versions of the natural landscape of the peninsula that aspired to form judgments, assessments, laws, and working theories about the phenomena of contours and territories. The observer had to spend time, not merely pass through, and needed historical and cultural data.

Terán found himself at home among the intellectuals at the Instituto Libre de Enseñanza in Madrid where French and German geographers influenced the collection of urban profiles of the city. He could compile these, registering the "*pulso dinámico*" ["dynamic pulse"] (Gómez Mendoza 24) in cuadrants and even street-by-street. And like Ortega, Terán found studies on urban geography—geo-demography—executed by scientists in France[5] and Germany of great use in his own work on Toledo, Baja Andalucía, Calatayud, or the towns of Aragón, and Sigüenza. In particular, the relationship between natural phenomena and cultural constructs implied the notions of process and change effected by human beings on their environment. That created a panorama of explanations tied to specific geographies that Terán could postulate and observe. Josefina Gómez Mendoza points out Terán's avoidance of Blanchard's geographical determinism by grounding his studies in visiting concrete sites rather than generalizing categories. When he wrote that "la ciudad es la transformación más radical de las naturaleza" ["the city represents the most radical transformation of nature"] (Gómez Mendoza 16), that observation signals a confluence of situation and site noted by the geographer and cartographer. A particular use of the terrain, with specific historical features, precludes the declaration of an inevitably determined human landscape. Far from the internal dreamscapes of Dalí, Terán's observed landscapes reflected historical and cultural interactions between human forces and inherited geographies.

Immigration to cities such as Madrid, the demographics of population change, and the interrelation of inhabitants and environment fill Terán's notebooks on Spain. To this he adds firsthand accounts of North Africa, an area not of prime interest to Spanish geographers of the time. He was as observant of the flora of the lands that Iberian populations chose to settle as were the early modern botanists and naturalists of the Spanish empire, but not for reasons of commerce or profit. Rather, Terán's details of rural life and urban development chart the possibilities of knowledge about human cultural and social adaptation, and about the history of geographical thought, in terms of a past and present that might inform future possibilities. Vivid descriptions and illustrations of Spanish geographical features done by Terán cemented the scientific basis and reputation of the journal *Estudios Geográficos* [*Geographical Studies*] through empirical observation complemented by further archival, historical and cultural research. No single aspect of these theoretical concerns was enough to stand on its own. Instead, Terán brought them together in a portrait of change across centuries. With them, he formed a new concept of "*paisaje*" ["landscape"] that went beyond occupying a space to the production of space by those who find, inhabit, modify, and evolve cultures within certain geographical areas. Terán wrote of geography as a study of the problems that arose from the "*instalación de los hombres sobre el haz de la tierra, su expansion en grupos crecientes por su número y densidad hacia horizontes de agrandada amplitud y su acomodación a espacios que organizan y componen con arreglo a dispositivos técnicos de progresiva eficacia y a exigencias y pautas culturales de ascendente valor y significación*" ["establishment of men on the face of the earth, their expansion in increasing groups, in both number and density, toward constantly expanding horizons and their accommodation to spaces that they organize and arrange with progressively efficient technical tools in response to cultural demands and models of increasing value and significance"] (Terán, "Una ética de la conservación" 377). What he termed the "*hominización del planeta*" (Terán, "Una ética de la conservación" 377) or the filling of the planet with human life called on the potential capacities of human beings as *Homo sapiens*, bearers of knowledge, as much as the exercise of their abilities to work in communities and as constructors of habitats. Terán rejected the passive accommodation of humans to their habitat in favor of the humanistic and scientific enterprises of building societies even beyond the observable limits of natural boundaries. Contrary to Dalí's internal, personal vision of the world, Terán grounded his explorations and notebooks in as thorough an observation of the world that he could accomplish.

As much as he was considered a geographer, Terán was an expert in "urbanism" (Gómez Mendoza 25), with its radical transformation and modernization of the Spanish landscape. This took place in particular across the decades between the 1940s and 1970s, at the time when Dalí was exploring his nuclear mysticism as a catalyst for art. Martínez de Pison and Ortega Cantero write that Terán was a man embedded firmly in the issues of his era: "No fue un geógrafo 'especializado,' recluido en su terreno y ajeno a lo que sucedía fuera de él, sino, al contrario, un geógrafo abierto a las inquietudes de su tiempo, directamente implicado en el panorama intelectual que le rodeaba" ["He wasn't a 'specialized' kind of geographer, secluded in his field and immune to what happened outside, but rather, on the contrary, a geographer open to the preoccupations of his time, directly involved in the intellectual panorama that surrounded him"] (9). For Terán, as for Benjamin, the layout and the districts of the cityscape was the prime object of visual interest for the scrutinizer of modernity. Observation continued to drive both Terán and Benjamin, even as others (such as Dalí) purported to question its value. Yet Benjamin's animation of the stones and spaces of Naples, Marseilles, or Moscow to awaken historical remnants from dream states into dynamic tension with the present deals with the ruins of temporality in ways other than Terán's more scientific empiricism. For Terán, intuition was allied with the technical apparatus available to discover "una ciencia de realidades concertas y visibles" ["a science of concrete, visible realities"] (Terán, "Una ética de la conservación" 381). Walking among the constructs and their ruins could reveal the disjunctions between the dreams of dominating a landscape and failure to do so productively. He did not end up lapsing into nostalgic paeans to lost values or divine inspirations. Instead, Terán began with the notion of the human being as an agent of organizing nature who in modern times has been afforded new and useful techniques and tools. He needed to observe how that had been done, and how it might be done differently

Benjamin's gaze evokes a response in the object of his vision—the "coming alive [that] is what Benjamin in his later work on photography will call its 'aura'" (Indyk 1). For his part, Terán methodically analyzes "*esta gran aventura de la urbanización*" ["this great urbanizing adventure"] ("Prólogo" 11) with careful, documented steps to search out the give and take between human cultures and how they affect geographies. Again like Benjamin, Terán does not posit a passive landscape and an active observer but an "*interrelación*" ("Prólogo" 11) or interpenetration akin to the "porosity" of characters and landscapes confronted by the critical eye whose role is to evoke time through the trace objects left behind. Benjamin's focus on the porous city—its crumbling walls, ancient buildings, rooms, interiors, facades, ironwork, marketplaces, cafés—as the locus of modern capital development, along with the resultant dehumanizing human experience, was more contradictory than Terán's. Benjamin saw in the city "the beautiful and the bestial," "exhilaration and hope, . . . revulsion and despair" (Gilloch, *Myth and Metropolis* 1). While Benjamin was never satisfied with the city as the site of a lasting, contented future, Terán's historical view saw Spanish cities as promises without set shapes, adaptations of landscapes and possibilities. The geographer's work on population density, concentric spheres of urban settlement, and evolving modification and transformation of the local topography as the "nucleus" of continual change, was rich in ideas as to how the "Gran Madrid" might look in the future but without a rigid plan for that development. Both find modern cityscapes as unresolved—and perhaps unresolvable—tensions among inhabitants, landscapes, and economics built on the ruins of previous conceptions of the city. And both are challenged to set foot on the streets in search of personal mappings of urban physiologies (what we have seen Benjamin call "physiognomies" [Gilloch, *Myth and Metropolis* 6]). The observation of social activities taking place along avenues and boulevards filled with emigrants to the attractions of modernity produced Benjamin's words and Terán's diagrams.

Terán too sustained the need to supplement and explain mere visible phenomena with patterns, theories, and the complex and contradictory relationship between places and inhabitants. This is evident in titles such as "El desarrollo metropolitano de Madrid: sus repercusiones geodemográficas" ["The Metropolitan Development of Madrid:

Geo-Demographic Repercussions"]). Yet he does not use language supported by images but instead a dialogue between language and image, akin to what Benjamin wrote of as "a dialectical optic" (Gilloch, *Myth and Metropolis* 7) of the everyday. Benjamin's beggars of Naples, bakeries of Moscow, or shopping arcades of Paris reinforce his theme of the fragmentation of city life. For Terán, the noise of the city ebbed and flowed, reflecting the immediate and local substrate in intimate contact with the inhabitants of a neighborhood, a street, a riverside, a hill, or the parliamentary power of the government. Terán observed the effects of topography on daily life, whereas Benjamin's city dwellers have all but obliterated the ground on which they have built and covered it with new merchandise, new social relations, and new commodities. Both are surveyors of the modern world that has overtaken the natural world in many and varied guises.

In Terán's writings, geography is not something set in stone but produced across time by generations of inhabitants. It is not *"inscrita en el suelo, la hacen los hombres contando con éste y a veces en contra de éste, pero lo que los hombres hacen sobre el suelo no es puro artificio, es geografía con el mismo derecho que la que se realiza al dictado de las condiciones físico-naturales"* ["inscribed in the ground or soil, but instead something men do depending on the solid ground and sometimes even doing harm to it, but what men do on the ground is not pure artifice, it is geography executed with the very same right of using the attributes of any natural physical conditions of terrain"] (Terán, "Prólogo" 9). Like Ortega's conception of human life or cultural geography, the world is something not inherited as is but built by individuals and communities. Terán finds in the natural resources of the earth the raw materials for any variety of constructs and relationships, not just a limited set of options. The maps he drew from his experience on the ground are merely the starting point for a representation of the "ruins" or semblances of human culture, the starting points not of *"ilustración"* ["illustrating"] but *"conceptualización e ideación"* ["the conceptualizing and designing of ideas"] ("Prólogo" 11). As much as Cajal could accurately observe phenomena, he then had to find how to communicate knowledge about them. Concepts and ideas contributed to development, modernization, and a different vision of Spain. Terán's notion of the superimposition of topographic layers and Benjamin's aura of the stratified ruin come together in the eye of both observers as archaeologists. The two coincide in envisioning the landscapes of human activity as layers of construction, of structures, and of remnants of past dreams. Whereas Benjamin stressed the "illusory and deceptive vision of the past" (Gilloch, *Myth and Metropolis* 13), Terán sought that *"pulso dinámico"* ["dynamic pulse"] (Gómez Mendoza 24) that anchored the foundations of the present but did not limit them. Benjamin held out hope for the eye of the observer to be radically altered by a spark of illumination. Terán accumulated the evidence for similar understanding.

Terán's procedure was to sketch out the objectives of the study. This would be followed by defining the boundaries and demarcation of spaces under consideration, the methodologies to be deployed, and the theories corresponding to a confluence of administrative records, measurement with the *cuadrícula* (diagonal and not always rectilinear), and reconsideration of the relativity between time and distance to urban centers. Following this consideration of geographical, logistical, and mobile details, Terán tracked the evolution of space and population in conjunction with one another, the density and distribution of the population concentrations across the region, and deviations in expected versus actual results. The density of a city was only the starting point for his observations, with abundant metaphors appearing for observed discontinuities in modern development. Terán described configurations of the city as *"como los granos de una granada cortada por la mitad"* ["like the seeds of a pomegranate cut down the middle"], *"de estructura tentacular o nebulosa"* ["of tentacular or nebulous structure"], and *"en manchas discontinuas, porosa y abierta"* ["with irregular, discontinuous concentrations, porous and open"] (Gómez Mendoza 24–25). Such details invite the participation of the observer formulate images, calculate shifts and changes in development, and comprehend science through metaphor. Terán's observations also solicit memories, comparisons, and multiple perspectives on the visible traces of the forces of modernity.

Like snapshots, Benjamin's short narrative fragments—the *Denkbilder* or linguistic capsules—captured discrete moments, places, and phenomena in the life of the urban labyrinth. But they are often "severed from a wider context" (Gilloch *Myth and Metropolis* 35) and live on as fragmented afterimages in the project on arcades and elsewhere. As Cajal built on the laboratory work of Golgi and perfected his stain to study the nervous system, so Dalí, Benjamin, and Terán added to the general public discourse on science with their acceptance or repudiation of inherited systems, images, and values. Terán walked the fields and roads of the Iberian Peninsula, looking at the visible surfaces of present life for traces of the intersection of past and present—what Benjamin had done in Naples, Moscow, and Marseilles—as emblematic visible histories of human adaptation. In each case, the porosity between phenomena and image, whether in the interpretation of photographs or the strata of human constructions, indicates not just the accumulations of modernity's dreams but also the increased role of the observer in making sense of those landscapes. Cajal had to learn to read and interpret what Golgi's stain revealed to him. Terán had to decipher what he as educated traveler could see, and what it implied. Dalí used scientific discoveries and innovations to feed his personal narratives. The discourse of science—from objectivity to observation, from experiment to theory—characterized the language of each and produced public figures that often attained cult status.

A Last Look at Observation

Observation has had a varied history, from the early modern era of imperial discovery and cataloging expeditions of naturalists and medical personnel, through the development of the microscope to make visible what was hidden from the naked eye. This trajectory includes the investment of the subject in interpreting images through a science of the eye, from nineteenth-century curiosity cabinets to the 1839 daguerreotype, from charting the urbanization of Castilla to a distrust of accurate vision after wartime, from Einstein's studies on relativity to Ortega's retinal vision, from Benjamin's porosity of surfaces that reveals an archaeology of meanings to color photography that could practically duplicate—even enhance—images. All of these situations and moments document the pervasive influence of visuality across the scientific professions as well as the arts, in ways both subtle and overt.

Activities related to the science of observation were encouraged by cultures that had learned to find value in the more detailed, the invisible made clear, the collection of knowledge about the world, and the commercial effect of what had been discovered. Observation would be held in even higher regard when it could be converted into invention and innovation. The results of observation were both intellectual and financial, but they were also iconic evidence of the modern. In times of expansion, there was the need for what Daston and Lunbeck refer to as a "calibration of the eyes" (Daston and Lunbeck 369) of Spanish explorers and colonial subjects for the collection and recording of data related to the natural world, as well as to investment and wealth. The circulation of goods and records of what had been seen linked center and periphery through communities of commodities that could be exchanged and circulated. One need think only of silver and gold, tobacco and chocolate, vanilla and corn to find the riches embedded in colonial enterprises of the discovery of flora and fauna. A subsequent contraction of power after 1898 turned the Spanish nation's gaze inward in self-observation, a reconsideration of the peninsula's resources and dreams, and a gradual romance with the visible devices of modernity that gave Spanish society the look of belonging to a world moving forward. Included among the inventions consumed by the Spanish public were photographs and photographic equipment, stereoscopes, dioramas, and kaleidoscopes. All were products of industrial techniques feeding into the consumer market that had established them as valuable. From approximately 1915 on, these objects were the portable and exchangeable signs of modernity. Portability and speed disrupted the static perception that had been a mainstay of scientific observation, and launched images and objects into greater circulation.

Decades later, in the twentieth century the relativization of the position of the observer, as well as technology based on mirrors and multiple lenses, would produce the divergence between Salvador Dalí and Manuel de Terán regarding the value of optical devices. Steadfast reliance by Terán on the surveyor's tools added to the depth of the human component of topography. The ultimate draughtsman, Dalí nevertheless rejected scientific measurement in favor of questioning the reliability of vision and the perception of space, time, and observation. Visual phenomena such as optical illusions that challenged the accuracy of what could be viewed, anamorphic images that held a surprise for the eye, and stereoscopy were the artist's weapons to challenge objectivity and propose instead irrational observation.

Spain in the nineteenth and early twentieth centuries was not a society remote from modern European communities of observation but rather in conversation with many of them, particularly in the wake of the soul-searching that followed the nation's defeat in the Spanish-American War. Conjecture about the future of Spain took the form of philosophical essays and reconsiderations of several possible paths. The first alternative looked toward the past and provided the resurrection of historical ruins, now made into monuments to be reconsecrated. A poetics of the dormant and increasingly depopulated countryside was extolled by some. The second alternative encompassed snapshots of metropolitan life that could capture the new rise of capitalism in boulevards, photographic labs, technological advances, and public personas of the scientist in what Walter Benjamin called "the theater of the new"

(Gilloch, *Myth and Metropolis* 41) found in modern Madrid. Benjamin's analogy of 1920s Moscow as "one giant scientific experiment" (Gilloch, *Myth and Metropolis* 41) with old and new confronting one another at every turn in terms of furniture, organizations, apartment houses, factories, laboratories, offices, and even the perception of time and space are hallmarks of Madrid as well. The figure of the flaneur or open-eyed observer, wandering amid the debris of modernity equates well with the figure of the photographer out among the teeming streets of the city, eyes sweeping the panorama of life. Benjamin referred to the constant movement of the street scene a vision "shot for a film" (Gilloch, *Myth and Metropolis* 41), with time reconfigured by participant and observer alike. His aim was to awaken the eye, just as Dalí and Buñuel had done with a quick slash in the opening scene of *Un chien andalou.* Cajal's memoirs capture his arrival in Madrid as a tension of opposing forces: scientific laboratory by day, sessions in the dark room by night.

As a man of arts and sciences, medical man and soldier, and a traveler in both Europe and America, Santiago Ramón y Cajal lived during a time of great advancement in image making. His public persona as a dedicated observer and experimenter humanized the scientist and professionalized the observer as covalent parts of Spanish culture. Cajal offered scientists and the general Spanish public an iconic figure of the modern, preserving a family structure of tradition but integrating that with a dedication to his scientific work. The family was part of that world, and its faces were to be studied as carefully as neurons were. Cajal dedicated a good amount of time to documenting his wife and children, his siblings and colleagues, and the landscapes of Aragón and Castilla that were changing with the arrival of modernity. That same cultural shift took Cajal out of Petilla de Aragón to Zaragoza, then to Madrid. In fact, one intuits the need for cosmopolitan contact with other scientists from Cajal's early days, during his stint in Cuba. His photographs of Petilla and other regional landmarks comprise albums of a disappearing geography, one that years later Manuel de Terán would walk, survey, and study. Transit—trains, electric trolleys, cars, and roads—sped up both the time associated with this change as well as the time an observer had to take stock of its ramifications. New ways of looking in the lab were accompanied by reconsiderations of time outside it and how to document them.

Cajal witnessed modernization in many guises: he recorded the streets of Madrid, the development of the university, the hawkers and

kiosks of avenues under construction, the development of water distribution, the burgeoning of the cinema and the phonograph, and
the critical importance of himself as observer. What Michael Ugarte
describes as "a city in a liminal stage of development" (2), in 1900
Madrid presented Cajal the opportunity to further his experiments
among more colleagues than he would have found in other cities.
Youthful excursions into the countryside, his convalescence from
malaria and tuberculosis in the rural hills, and the camaraderie of the
Gaster Club on weekend treks, provide images of tension between
the relationships of the past and the harried life in Madrid. Not just
his own age but changing relations with time are visibly imprinted
on the photographic images Cajal collected in his albums. His later
writings reveal a sentiment of loss as he saw the speed of culture, and
he feared the eclipse of his own faculties. His latent jealousy of young
scientists born at a time of rapid social change who would bear witness to many new inventions and discoveries that he would not live
to see filled his memoirs as the two distinct measures of time—slowing down, pressing ahead—came together. If his early days were filled
with acquiring medical knowledge, Cajal also captured anthropological
scenes of Valencian culture taken in celebrations with coworkers in
the woods and fields. Cajal's photographs suggest a rich and complex
counterpoint between experimentation and observation, a synthesis of
science with a study of the society around him.

There is a little of the pictorialist that lingers on in the man
who spent so much time perfecting the science of lenses and light,
but in the end Cajal is a figure that embodies the nation's own transition. Einstein joined Cajal as a guiding light and cult hero for the
general public in a Spanish culture focused on observation. Einstein's
ideas were promoted and consumed by all levels of society, creating
a sense of participation through the purchase of books, explanatory
guides, cameras, and other objects associated with science. The Spanish
fascination with science and its heroes would last through the prewar
years of the 1920s and 1930s, resurfacing in the 1940s, 1950s, and
1960s in different ways.

On the one hand, Manuel de Terán could not wait to take up
the urban regeneration of Madrid as a cause in the 1950s and 1960s,
noting the obsolescence of some projects and the urgent need for
the cultivation of new ideas of space. His vision of population centers
as "redes" or nets of commercial and residential centers linked with

natural resources and green spaces relied on the actual physical observation of the areas in question. Rather than rely on maps or outdated studies, Terán trusted his direct observation—the requirement to get out and "*andar y ver*" ("walk and observe")—for providing information that might offer alternatives to the uninhabitable city that "*es la mutación más radical a la que puede ser sometido el paisaje natural*" ["is the most radical mutation to which the natural landscape can be subjected"] (Terán, *Ciudades españolas* 387). There could be no inspiration for change without direct observation. Even those cities that had for a longer period of time retained their archaic or preindustrial structures in some way experienced the changing forces of modernity. The project that remained, wrote Terán, was to rehumanize the economic promise of urban life since a tendency toward the agglomeration of cities had been incontrovertible. He examined how to seek survivals of what Benjamin called the internalized social dreamscapes that brought inhabitants to those sites to reconnect "la comunión del hombre con la tierra" ["the communion between man and land"] (Terán, *Ciudades españolas* 391).

On the other hand, the inventory of human physiology mapped during the nineteenth century, along with Cajal's study of the cerebral cortex, collided with a shift in the value of observation. Associated with the work on lenses by French physicist Augustine Jean Fresnel, a change in the notion of the characteristics of light had occurred. With light as the invariable foundation for observation, new relationships between light and color, light and electricity, and the pulsation of waves in the transmission of light took the basic notions of vision and light into more complex territory. Dalí was cognizant of those experiments, and they suggested ways he could integrate their science into his art. Obviously, Dalí's opposition to the unquestioned reliability of the sight did not intersect with Terán's measurement of topographic spaces. Yet each assimilated an emphasis on vision that connected Cajal with Ortega, Vizuete with Einstein, and Benjamin with European modernity. The continued relevance of the lens, vision, the instability of the modern observer, the effects of light and electromagnetic waves, and the establishing of the scientist at the core of systems of knowledge are evident across the decades explored in Spain's changing cultural landscape.

Like Cajal before him, Terán was one of few Spaniards who participated fully in the intellectual and cultural horizon of the moment. In his work, science and cultural geography converge around what

Ortega had proposed as the conceptual difference between *medio* and *paisaje*. The first corresponds to Terán's liberating "posibilismo" or potential for interaction between a place and its inhabitants, rather than the "determinismo" (Martínez de Pisón and Ortega Cantero 183) or rigid causality of a geography that held sway over humans. Human circumstances, as much for Ortega as for Terán, were the raw material for self-construction and adaptation, not a life sentence. Science could function at the service of societies which learned from the strata of the past; "la utilización del medio por el hombre, el diálogo que con él entabla, los vínculos que con él anuda, dependen, en última instancia, de la idea que se hace de su posición en el mundo . . . en función de sus necesidades y de los fines que se propone, sujeta, pues, a variaciones históricas" ["the utilization of the environment by man, the dialogue he establishes with it, the ties he creates, depend, ultimately, on his idea of his own position in the world . . . according to his needs and the goals he proposes, subject, of course, to historical variants"] (Martínez de Pisón and Ortega Cantero 183). The echoes of Ortega are evident. Behind Terán's walking, observing, and recording, the observations of Humboldt stood as close by as the scientific innovations of mid-century. This included hypotheses about tectonic plate movement, the protection of nature, and the conclusion that geography was fundamentally *"una ciencia del paisaje"* ["a science of landscape"] (Martínez de Pisón and Ortega Cantero 184). All phenomena occurring in a region join the mobile observer in the production of one's *"circunstancias."*

"Andar y ver," a phrase shared by Ortega and Terán that was also about the philosopher's gaze, connected the geographical horizon with the intellectual one. The concrete aspects of a determined landscape combined with the mobility of the subject, with a perspective drawn to the luminous and the illuminated, to the effects of light on space. Landscape stopped being a still life (*naturaleza muerta*) and was brought to life through observation by the traveler, scientist, artist, or wandering scientist. Of course, optics mediated all encounters between observers and their environs: "La sujetualidad del paisaje es el resultado de la disolución de la inercia que se capta como naturaleza" ["Landscape as a subject is the result of the dissolution of the inertia that is usually captured as 'nature'"] (Paredes Martín 182). The sleepwalker does not see outside the dream; something must cut through the gaze to permit a conscious connection.

Benjamin's momentary spark of comprehension of landscapes and cityscapes—natural and cultural settings—marks a turn from the habitual to the engaged, from a crisis of the eye to a new optics. The material artifacts that humans produce to give form to a landscape, or the urban spaces humans construct, turn cities into laboratories of modernity for him as they do for Cajal and Ortega. The social experiments that address human problems in visible public spaces become experiences for commentary, whether affirmational or refutational. Some of those comments are contained in essays like those by Terán and Ortega whose optical regime underscored a close if fleeting union between the observer and the observed. Dalí struck a different chord with his paranoid-critical method that asserted the site of image making as the subconscious, not the awakened observer. In any case, the eye was the preeminent key that connected to the nervous system and the cognitive function of the brain, *"una especie de diccionario pictórico"* ["a type of pictorial dictionary"] (Cajal, *Mi infancia y juventud* 113). Even if Dalí's repertoire of images did not reflect conscious vision but an uncontrolled, uncensored, and subjective set of symbols sprung from the natural world but removed from mimetic identification with it, vision was still critical.

The principal object of concern—and a fundamental subject of public discourse—was modernity itself and what it might look like in Spain. The transformations of Madrid, the concerns of intellectual *tertulias*, the proliferation of photographic studios and portable image-making equipment, architectural planning, population density and water resources, and migration from provinces to cities all contributed to science becoming a public affair. Natural sciences that united empires and commercial observers gave way to laboratories dedicated to treating tropical diseases. The building of a royal court ceded to nation-building and urban development. Botany shifted to machinery, engineering yielded to neurology, physics, and astronomy under the aegis of Cajal and Einstein. Like geomorphology itself—the science of landscape—Spain's fostering a look of modernity was "built very clearly on the work of the past" (Goudie and Viles 14). Technological innovations contributed to new types of knowledge whose pace of assimilation accelerated with the temporal impetus of modern times, and to the dissemination of that knowledge across greater sectors of Spanish society. Through books and pamphlets, Pelayo Vizuete could promulgate the ideas of new physics to more readers using the language of metaphor that Benjamin called the space of ruinous traces (language uniting past and present, the known and the unknown). Ortega could address Spanish readers through the mass media, recording his eyewitness accounts of both provinces and capitals of Spain, evoking signs of change (crumbling factories, cloistered monasteries, railway tracks) in the linguistic conventions of a vocabulary related to sight and observation.

To reach his audience, Ortega employed similar metaphorical language to trace the visible reflection of interior processes. He began with the eyes as a proscenium arch that surrounds the drama of life observed on the face as if it were played out on a stage for spectators to observe. The comparison is revealing:

> Bajo el arco de las cejas, como tras de la boca del escenario, párpados, esclerótica, pupila, iris integran una maravillosa compañía de teatro, que representa maravillosamente el drama y la comedia de dentro. Es inconcebible que no se haya hecho aún—que yo sepa—el vocabulario de la mirada, que no se hayan clasificado los modos de ella. La mirada recta y la de través, la mirada prensil que llega al objeto y queda en él agarrada, y la mirada blanda que resbala sobre su forma sin prenderla. . . . La mirada indiferente, la intensa, la vaga . . . Se comprende que sea la mirada, de las porciones visibles del cuerpo, la más rica en poder expresivo. En el aparato ocular intervienen el mayor número de músculos pequeños y sumamente sensibles, que obedecen a las menores presiones del ser íntimo.

> [Under the eyebrows' arch, just as behind the theatrical stage, eyelids, sclera, pupil, iris form a marvelous theater company that represents, also marvelously, the drama and comedy of what lies inside. It is inconceivable that no one—at least as far as I know—has composed a vocabulary of the gaze, that no one has classified what kinds of gazes there are. Straight on and sideways glance, the prehensile gaze that fuses with the object and the soft gaze that slides over a surface without ever getting in. . . . The indifferent gaze, or intense, or vague . . . It is entirely understandable that the gaze of the eye, of all the visible parts of the body, is the

richest in expressive power. The ocular apparatus is made up
of the greatest number of small and most sensitive muscles
which respond to even the tiniest pressures of one's interior
being.] (Ortega, "Sobre la expresión" 691–92)

Ortega mixed physiological processes and philosophical responses in
this description of how the eye functions to apprehend the world,
how it leads to knowledge and interpretation, and how the observer
and the observed intermingle. Interior psychological and biological life
and external social life coalesce in the eye.

As an actor on the stage of Spain's social and economic romance
with science, Ortega propounded the ways in which radically new
technologies and innovations could respond to human life rather than
be collected as evidence of integration into a European theater of the
modern. He addressed the infiltration of Einstein's theories into Spanish culture, the popularizing of scientific figures, the democratizing of
art and culture, and the effects of technology on the perception of
the passage of time. So it was natural that he would articulate many
of his publications around the organ of sight as a portal of perception. The observer—*El Espectador*—was not a passive onlooker but a
traveler on the two-way street of culture where science had become
both medium and message, both utopia and dystopia.

After addressing the history of challenges between mathematics
and science, Ortega sums up the obligation to keep up with the latest
debates in physics: "La filosofía misma, que necesita tan pocas cosas, ha
menester, sin remisión, de la física para poder ser lo contrario de ella,
que es su misión" ["For philosophy itself, which requires so few things,
physics is indispensable so that philosophy can be its opposite, which is
its mission"] ("Bronca" 153). For Ortega, the fact that so much talk of
the universe became part of public discourse indicated that physics had
gone beyond the limits of *"lo observable"* ("Bronca" 157) into the realm
of an imagined universe. He concluded that the sum of all imaginary
worlds would have to come together in order to be compared with
observed facts. As strange as it might sound, the art of Dalí did just that.
He intermingled sleeping and waking worlds, and natural and psychic
phenomena into constellations of simultaneous forces. Yet Dalí converted
these phenomena into cash, not theoretical models for analysis.

At eighty, Cajal wrote insistently about changes in the world
wrought by *"el tiempo, el progreso y la moda"* ["time, progress, and fashion

or fads"] ("La fotografía" 65) that he examined through shifts in the Spanish language and customs. He lamented the unrecognizability of towns he knew in the past, and that this disjunction in what he could no longer see there cut him off from his memories. Renovation and progress, and "*la piqueta demoledora*" ["the relentless pickax"] ("La fotografía" 66) that demolished the past, rivers of automobiles that placed walkers in danger, were the observable machinery of modernity. Cajal found it acceptable to champion modernization in the laboratory, but less so in the society around him.

For Benjamin, Einstein, Vizuete, Dalí, Terán, and Ortega, the modern world of the twentieth century accommodated technological accomplishments and scientific precision, precipitated obsolescence, and produced radical shifts in the value of ideas and commodities. Modernity was not so much a singular goal as a complex and uneven process, the subject of radical art and the object that artists prized above all. Scientific modernity gave a semblance of the new to Spanish culture and made it—like the walls of Naples, Moscow, or Marseilles were for Benjamin and the landscapes of Madrid, El Escorial, Salamanca, Valladolid, or Oviedo were for Ortega and Terán—a porous space of permanent fusion and experimentation. For them, modern Spain was a project in the works worthy of observation.

Notes

Introduction

1. As Holton indicates, "by the early part of the nineteenth century, a long-simmering rebellion came to a boil against the hegemony of Euclidean geometry and especially its so-called fifth axiom; that one implies, as our schoolbooks clumsily state it, that through a point next to a straight line only one line can be drawn that is parallel to it, both of them intersecting only at infinity" (2001).

Much as Pelayo Vizuete made numerous attempts to educate a popular audience about Einstein's theories, Poincaré represented notions about debates in higher geometry. He was, like Salvador Dalí, a bridge between scientists and artists.

Chapter One

1. Both sides in this conflict—what Cirillo calls "a *little* war with *big* consequences" (1)—misread the topography of Cuba, the political events previous to the start of war, and the bacteriological warnings of germ theory generally already available. Indifferent to sanitary conditions, officers and their men suffered equally in the armies of both Spain and the United States. Decimated by typhoid fever, malaria, and other tropical ailments, they even took some of these maladies home with them. In the 1898 conflict, the practical sciences did not do their job in preventing medical losses in great numbers.

2. Cajal exemplifies the drive to science in several ways. Carl Gustav Hempel includes two principal factors in the enduring human attributes and concerns of the empirical sciences; he finds these to be the stimuli behind research efforts and investigations in general. Hempel concludes that the first concern is

of a practical nature. Man wants not only to survive in the world, but also to improve his strategic position in it. This makes it important for him to find reliable ways of foreseeing changes in his environment and, if possible, controlling them to his advantage. . . . The second basic motive for man's scientific quest is independent of such practical concerns; it lies in his sheer intellectual curiosity, in his deep and persistent desire to know and to understand himself and his world. (333)

It is evident that Ramón y Cajal responded to the second impetus, the drive to acquire understanding of the world, while the practical applications of his work were obvious albeit with less direct influence over him. His work with the chemicals of photographic processes, and with the color quality of images, was driven more by inquisitiveness than by lucrative patents or proprietary holdings. He set out to clarify the improvements made in color photography so that others might benefit from better images, whether of cells or of family members. One might add to these two dimensions another in the case of Cajal: his fervent desire—"passion" as Laín Entralgo and Albarracín see it—to cast a positive light on his native Spain and the persistence of science there despite all financial shortcomings and lack of official support. This drove him as much as the first two qualities did. Dr. Juan A. de Carlos, currently in charge of the vast scientific and intellectual legacy included in the Museo Cajal (at the Instituto Cajal in Madrid), has characterized Cajal's brilliant career within the scientific community in Spain as responding to a variety of factors. In a recent interview, de Carlos writes in answer to a question:

> ¿Había vida en la Ciencia Española? Evidentemente sí, pero también había importantes lagunas que Cajal supo aprovechar. Por ejemplo, en Histología del Sistema Nervioso, fue pionero en su tiempo, dada la dificultad existente en la impregnación de dicho tejido, por lo que se encontraba prácticamente sin estudiar. En España, desde luego no había nada. . . . Cajal fue capaz de aventajar a todas las escuelas histológicas europeas. . . . Se impone, a mi entender, su genio con el que es capaz de tomar ventaja en las diversas circunstancias que se le van presentando. (desdeelexilio.com 3)

> [You ask me was there life in Spanish science? It is evident that there was, but important lacunae also existed and those Cajal knew how to take advantage of. For example, in the histology of the nervous system, he was an early pioneer, and given the difficulty of staining those tissues, they remained practically unstudied. In

Spain, naturally there was nothing. . . . Cajal was able to get the jump on all the schools of histology in Europe. . . . What stands out, in my understanding, is his genius for taking advantage of the diverse circumstances that came along.]

Not an intellectual of spontaneous generation, Cajal is indeed the most prominent Spanish scientist—de Carlos considers him more cited than either Einstein or Darwin in their respective fields—among figures such as "Isaac Peral (1865–1895), inventor del submarino, . . . Juan de la Cierva (1895–1936), inventor del autogiro, o . . . Leonardo Torres-Quevedo (1852–1936), uno de los inventores más prolíficos que ha tenido nuestro país" ["Isaac Peral (1865–1895), inventor of the submarine, Juan de la Cierva (1895–1936), inventor of the autogyro (a type of rotorcraft similar to the helicopter), and Leonardo Torres-Quevedo (1852–1936), one of the most prolific inventors Spain has ever had"] (desdeelexilio 2). It is obvious that scientific studies in Spain were present in diverse fields, from civil engineering to marine biology, from aircraft engineering to electromagnetics and robotics, from analogue calculating machines to pathology. These and all subsequent quotes from Spanish are my translations unless otherwise noted.

3. Gerardo F. Kurtz examines the effects on Spanish culture of the translation of Daguerre's text:

En 1839–40 aparecerán en España tres traducciones del manual que hubo de producir Daguerre como parte del acuerdo que firmó con el estado francés y mediante el cual acepta dar a conocer libre de patente su invento del daguerrotipo. . . . El manual impreso, que en virtud del mencionado acuerdo debe producir Daguerre para explicar en profundidad el proceso daguerrotípico, no vio la luz hasta principios del mes de septiembre de ese mismo año [1839]. (Kurtz 1996, 4)

[In 1839–40 three translations appeared of the manual that Daguerre was to produce as part of the agreement he signed with the French state, and in which he was to introduce his invention free of any patent. . . . The printed manual that according to the agreement would explain in detail the process of the daguerreotype did not appear until the beginning of September of that year (1839).]

The promise of a generous pension from the French government did not hinder Daguerre's anxiety for some commercial benefit from his invention. Kurtz continues:

De este manual existen en castellano tres versiones distintas, todas ellas traducciones con fecha de publicación de 1839. . . . Estas traducciones se deben a Eugenio de Ochoa, . . . Pedro Mata y Fontanet, . . . y a Juan María Pou y Camps en colaboración con Joaquín Hysern y Molleras. . . . Esta última traducción es especialmente interesante desde el punto de vista histórico, editorial y fotográfico por contener numerosas *notas de traductor* en las que se hacen muy extensas y autorizadas referencias a la técnica, práctica y cicunstancias teóricas del proceso daguerrotípico, incluso describiéndose en varias ocasiones los procederes concretos de la toma de vistas daguerrotípicas específicas. (Kurtz 1996, 5–6)

[There are three different versions in Castilian (Spanish) of this manual, all published with the date of 1839. . . . These translations were done by Eugenio de Ochoa, . . . Pedro Mata y Fontanet, . . . and Juan María Pou y Camps in collaboration with Joaquín Hysern y Molleras. . . . The last of the three translations is especially interesting from the historical, publishing, and photographic point of view since it contains numerous *translator's notes* in which extensive, authorized references to technique, practice, and theories of the daguerreotype process are made, including several concrete references to the procedures related to the taking of specific daguerreotype scenes.]

Given the brevity of the other two versions, one might surmise that the last of the three would be of most professional interest to the type of detailed, and very scientific, photographer Cajal was. The notes of the translator of one of the editions of the text, one whose quest for linguistic accuracy would parallel Cajal's quest for objectivity in the lab, as well as the detailed instructions for chemical processes, appealed to the man of science whose "scientific quest retains a space for the play of the imagination" (Pratt 12). Pratt's important study refers to the rhetoric of Cajal's scientific and literary writings, but his photographic work was equally rich in imaginative visual compositions and the promotion of innovation. Although Cajal "wants science to transcend its linguistic nature and refer transparently to the world of objects, [in his short stories] he demonstrates . . . that such transparency is impossible" (Pratt 100–101). One may strive for the "right seeing" invested in the language of empiricism as an accompaniment to the modern, but the translation of material objects into visual images, or into linguistic reports, is always predicated on a process that transliterates from one system of mapping into an entirely different system.

Chapter Two

1. Paul Martineau points out that the year 1839 established Daguerre's perfected equipment as the standard for photography, but his partnership with inventor Nicéphore Niépce began much earlier (6). With Niépce's death in 1833, the path was left open for Daguerre to refine and promote their process of creating images on silver-coated copper plates. As the chemical processes used in photography became more advanced, times for exposure and development were altered and a broader range of objects could be recorded.

2. Eugenio Portela and Amparo Soler examine the protagonists of the slow rise of experimental chemistry in Spain, even in the relative social truce after the revolution of 1868. They find two critical details that underscore an emphasis on the importation of knowledge and photographic processes into Spain rather than a vast original and sustained development: the lack of investment in laboratory equipment and the formation of teams rather than individual researchers (the latter being fairly impossible in Spain given the low numbers of men engaged in such activities). They write that toward the end of the nineteenth century, "La investigación química europea había pasado de ser una actividad personal a una labor de equipo, con investigadores profesionalizados ["Research in chemistry across Europe had changed from personal activity to team work, with professional researchers"] (Portela and Soler 101). The professionalizing of medicine in Spain was a given as were those dedicated to chemical work for industry and agriculture (vineyards), but academic chemistry did not sponsor investment in photography.

The same authors also signal an institutional foundation for Germany's domination of this area starting in the mid-nineteenth century, with industrial needs accompanied by equal investment in personnel and equipment in a new university model for such work (102). Cajal's manual on color photography's scientific underpinnings reveals his thorough knowledge of photographic processes learned from manuals imported from France and Germany, including his own translation of such manuals. Specifically, Cajal dedicated quite a bit of space in the volume to Gabriel Lippmann (a physicist awarded the 1908 Nobel Prize for his work in recording and preserving color images) as well as to Lumière, Neuhaus, Zenker, and Berthier-Ives. In fact, in 1907 one of Cajal's monographs of 1906 on the photochrome process of Lippmann was translated into German as a study of Lipppmann's structures in a manual on the chemistry and physics of photography (see Cajal, "Fotografía de los colores" 287–89). Cajal explained his enthusiasm for the translation of his own work into as many European languages as possible as a response to a pressing need. His work published in French, Italian, and German periodicals in those languages, confirming his conclusion that very few colleagues would be capable of reading scientific Spanish. To get his work known abroad, he was convinced he would have to get it translated.

As Klaus Biedermann explains, Lippmann used the wave nature of light to produce a phenomenon of interference in the formation of the photographic image. His focusing of light waves is captured by Cajal as the third of three processes of color reproduction (after those of Cros and Becquerel) that uses the work of Neuhaus and Zenker to work with the spaces between the colors of the spectrum to create equilibrium, intensity, and transparency in the colors reproduced (Cajal, "Fotografía de los colores" 287–88). A scholar of optics, astronomy, and seismology (Biedermann 5), Lippman also pioneered what would become three-dimensional imaging in the 1960s through his pursuit of "integral photography" (in work published in 1908 and accessible to Cajal). For Cajal, the interferential process using light waves was a mechanism whose theoretical bases were to be put to the practical test in the laboratory. The process of investigation related to human cells is equivalent to the research necessary on the reproduction of images. Both must be proven and, if possible, improved so that knowledge may be increased. It is clear that physics, chemistry, and photography all coalesced in the work of Cajal, if not as part of an extensive team that might have been supported elsewhere then at least as the model of scientific inquiry for future generations.

Chapter Three

1. Peter L. Galison's exciting study of Einstein and Poincaré, focusing on the confluences and differences of each one's contributions to a new vision of the physical world and to theories related to relativity and the conventions of measuring time, addresses the nineteenth century's study of waves of light and their perception by subjects from "moving frames of reference" (Galison and Burnett 2). While the notion of light waves and theoretical Physics seem remote from the harnessing of light for the illumination of cities, residences, and public streets, Galison's return to the train station of Einstein's hypothesis to unearth the relationships between "the order of theory itself [that would mirror] the order of the world" (Galison and Burnett 8) supports the idea that these are not separate realms but rather "[the bringing] of the abstract into the concrete" (Galison and Burnett 12). So the abstraction of color photography and its implementation in material technologies as Cajal sought does not fall outside the space of clocks, meridians of longitude (Poincaré), and the acquisition of knowledge through concrete products. Electricity at once literally cast light on dark alleyways and shed light on the movement of waves through a vacuum. What Galison beautifully refers to as "mingling machines and metaphysics" or "the nearness of things and thoughts" (11)

was evident in the advances of urban life and the debates that took place around them. Technology and physics were two sides of the same coin: "When Poincaré and Einstein looked into the details of electrical engineering, when they stared at generators, radios, and cables, they saw in them critical problems of physics and philosophy" (11–12). Everyday objects came about as a result of questions about the material universe, and technologies harnessed and deployed the discoveries.

Chapter Four

1. It is noteworthy that the "catastrophism" linked to Einstein's influence on thinking about dynamic systems was an object of fascination for the Spanish artist Salvador Dalí. French mathematician René Thom maintained a correspondence with Dalí related to the artist's anxieties arising from the atomic age and its destructive potential but also from what Dalí discerned as the potential for a reconsideration of the image in the light of such dramatic changes in equilibrium and perception.

2. Vizuete also published linguistic guides such as *Lecciones de árabe marroquí* [*Lessons on Moroccan Arabic*] and was an editor of and contributor to the *Diccionario Enciclopédico Hispano-Americano* [*Hispanoamerican Encyclopedic Dictionary*] published in Barcelona in 1907, 1908, and 1910.

3. Concerned with maintaining the social relevance of geography, and invested in the scientific accuracy as well as accessible style of his work, Terán was stopped from his research walks in Toledo by the outbreak of the war. No longer could the productive mutual understanding between nature and culture be observed with his own eyes, as wandering through the provinces became unadvisable or prohibited. One result was that this project was left incomplete. But another was a turn to Madrid as the object of his observation, a city in the geographic center of the peninsula, the battles, and the fight for modernity. Terán's dedication to "a geography 'alive and human,' a geography centered on man, . . . a geography able to 'make a home from things'" (García Ballesteros 11) required that he turn to previous observations from before wartime and explore the forces at work in the development of the city in the 1950s. He began, out of political necessity, with the Madrid of the Austrian monarchs.

4. Angel Cabo Alonso has studied not just the variety of writings by Terán—on cartography, history, sociology, demographics, commerce, conservation—but its stylistic components as well. Among them, he mentions Terán's work on the countryside from Tarragona to Santander and how Ortega y Gasset's talks in the Instituto Libre de Enseñanza and the Residencia de Estudiantes focused the young geographer's attention on "el respeto a las cosas, a

las múltiples cosas naturales o de creación humana, con sus nombres, formas y colores" ["a respect for material things, whether they be the multiplicity of natural objects or those of human creation, along with their names, forms, and colors"] (Cabo Alonso 143). Such an interest in the interrelationships of elements among which human beings live puts into relief both institutions and landmarks, food reserves like grain elevators and tools, the uses of habitational space and the development of factories, the fabrication of buildings or their demolition, often employing similes and metaphors to explain references in terms of known or more familiar images. With these linguistic tools, Terán and Ortega could communicate to a vaster audience. In the end, geography as a field put landscapes under the lens of scrutiny and turned from the metaphors of the Generation of 1898 (lost idylls, forgotten values, crushed utopias) to concrete spaces of development recorded in documents and archives, on maps, and in reports. Direct knowledge of any landscape, and eyewitness account of its details, negated the sublime romanticism of re-created spaces tinged with emotion. In place of this, Terán reflected on what he had actually seen, a dialogue between the observer and the observed in which "Todo el paisaje es pedagogo" ["All landscape is a teacher"] (Martínez de Pisón 129).

5. Terán cited the work of French geographer P. Gourou on "Determinismo y posibilismo" (Terán, "Una ética de conservación" 379) as opposing poles in which culture—he calls it "la civilización"—intervened to help inhabitants of a landscape progress beyond mere adaptation. Echoing the language of Ortega y Gasset, Terán defined civilization as "un repertorio de ideas y creencias, instituciones y usos, normas de conducta social, técnicas de trabajo material. Todo un complejo de creaciones espirituales y materiales, que tiene su reflejo en la manera como los grupos sociales hacen su instalación en un medio, lo interrogan" ["a repertoire of ideas and beliefs, of institutions and practices, of norms of social, and of techniques of material labor. (It is) an entire complex of spiritual and material creations that is reflected in the ways in which social groups settle into an environment, how they examine it"] (Terán, "Una ética de conservación" 381).

Works Cited

Achim, Miruna. "Science in Translation: The Commerce of Facts and Artifacts in the Transatlantic Spanish World." *Journal of Spanish Cultural Studies* 8.2 (July 2007): 107–115. Print.

Anes y Alvarez de Castrillón, Gonzalo. "Presentación." Manuel de Terán, *Ciudades españolas (Estudios de Geografía Urbana)*. Madrid: Real Academia de la Historia, 2004. 9–10. Print.

Antonio, Robert J., ed. *Marx and Modernity. Key Readings and Commentary.* Malden, MA: Blackwell, 2003. Print.

Ashworth, William B., Jr. "Natural History and the Emblematic World View." *Reappraisals of the Scientific Revolution.* Ed. David C. Lindberg and Robert S. Westman. Cambridge: Cambridge UP, 1990. 303–333. Print.

Bajac, Quentin. *The Invention of Photography.* London: Thames and Hudson, 2002. Print.

Barthes, Roland. *Camera Lucida.* New York: Hill and Wang, 1980. Print.

———. "Rhetoric of the Image." 1964. *Image Music Text.* Trans. and selection Stephen Heath. New York: Hill and Wang, 1978. 32–51.

Benjamin, Andrew. "Porosity at the Edge: Working Through Walter Benjamin's 'Naples.'" *Walter Benjamin and Architecture.* Ed. Gevork Hartoonian. London and New York: Routledge, 2010. 39–60. Print.

Benjamin, Andrew, and Charles Rice. "Walter Benjamin and the Architecture of Modernity." *Walter Benjamin and the Architecture of Modernity.* Ed. Andrew Benjamin and Charles Rice. Melbourne: re.press, 2009. 3–8. Print.

Benjamin, Walter. *Illuminations: Essays and Reflections.* Ed. and introduction, Hannah Arendt. Trans. Harry Zohn. New York: Schocken Books, 1968. Print.

———. *Moscow Diary.* Ed. Gary Smith. Trans. Richard Sieburth. Cambridge, MA: Harvard UP, 1986. Print.

———. *Reflections: Essays, Aphorisms, Autobiographical Writings.* Ed. and Introduction, Peter Demetz. Trans. Edmund Jephcott. New York: Schocken Books, 2007. Print.

————. "Surrealism: The Last Snapshot of the European Intelligentsia." *One-Way Street and Other Writings*. Trans. Edmund Jephcott and Kinglsey Shorter. London: Verso, 1985. 225–39. Print.

————. "Theses on the Philosophy of History." *Illuminations: Essays and Reflections*. Ed. and introduction, Hannah Arendt. Trans. Harry Zohn. New York: Schocken Books, 1968. 253–64. Print.

Biedermann, Klaus. "Lippmann's and Gabor's Revolutionary Approach to Imaging." http://nobelprize.org/nobel_prizes/physics/articles/bidermann/. Accessed July 2013. 1–7. Web.

Bleichmar, Daniela. "A Visible and Useful Empire: Visual Culture and Colonial Natural History in the Eighteenth-Century Spanish World." *Science in the Spanish and Portuguese Empires, 1500–1800*. Ed. Daniela Bleichmar, Paula De Vos, Kristine Huffine, and Kevin Sheehan. Palo Alto, CA: Stanford UP, 2009. 290–310. Print.

Blumenberg, Hans. "Light as a Metaphor for Truth At the Preliminary Stage of Philosophical Concept Formation." *Modernity and the Hegemony of Vision*. Ed. David Michael Levin. Berkeley: U of California P, 1993. 30–62. Print.

Brooke, John Hedley. *Science and Religion: Some Historical Perspectives*. Cambridge: Cambridge UP, 1991. Print.

Buchen, Lizzie. "May 29, 1919: A Major Eclipse, Relatively Speaking." 1–3 http://www.wired.com/thisdayintech/2009/05/dayintech_0529/. Print.

Buck-Morss, Susan. *The Dialectics of Seeing. Walter Benjamin and the Arcades Project*. Cambridge, MA and London: MIT Press, 1991. Print.

Cabo Alonso, Angel. "El campo español en los escritos del Profesor Teran." *Manuel de Terán Geógrafo 1904–1984*. Ed. Eduardo Martínez de Pisón and Nicolás Ortega Cantero. Madrid: Residencia de Estudiantes/Sociedad Estatal de Conmemoraciones Culturales. 137–59. Print.

Cantor, Geoffrey and John Hedley Brooke. Section III: "Having Designs on Nature." *Reconstructing Nature, 1995–1996*.139–55, 156–72, 173–90. http://www.giffordlectures.org/Browse.asp?PubID=TPRECN&Volu. Accessed July 2013. Web.

Cañizares-Esguerra, Jorge. "Introduction." *Science in the Spanish and Portuguese Empires, 1500–1800*. Eds. Daniela Bleichmar, Paula De Vos, Kristine Huffine, and Kevin Sheehan. Palo Alto, CA: Stanford UP, 2009. 1–5. Print.

Castro, Fernando de. "Presentación del libro 'La fotografía de los colores,' de Cajal." *Fotografía de los colores: bases científicas y reglas prácticas*. Zaragoza: Prames—Las tres sorores y Gobierno de Aragón, 1966. 5–12. Print.

Caygill, Howard. *Walter Benjamin: The Colour of Experience*. London and New York: Routledge, 1998. Print.

Charnon-Deutsch, Lou. *Hold That Pose: Visual Culture in the Late-Nineteenth-Century Spanish Periodical.* University Park: The Pennsylvania State UP, 2008. Print.

Cirillo, Vincent J. *Bullets and Bacilli: The Spanish-American War and Military Medicine.* New Brunswick, NJ: Rutgers UP, 2004. Print.

Clark, Stuart. *Vanities of the Eye: Vision in Early Modern European Culture.* Oxford: Oxford UP, 2007. Print.Clark, T. J. *The Painting of Modern Life: Paris in the Age of Manet and His Followers.* Princeton: UP, 1984. Print.

Crary, Jonathan. *Techniques of the Observer: On Vision and Modernity in the Nineteenth Century.* Cambridge, MA: MIT Press (An October book), 1992. Print.

Dalí, Salvador. *The Secret Life of Salvador Dalí.* Trans. Haakon M. Chevalier. New York: Dover, 1993. Print.

Darwin, Charles. *Evolutionary Writings.* Ed., Notes, and Introduction James A. Secord. Oxford: Oxford UP, 2008. Print.

Daston, Lorraine, and Elizabeth Lunbeck, eds. "Introduction: Observation Observed." *Histories of Scientific Observation.* Chicago and London: U of Chicago P, 2011. Print.

Daston, Lorraine, and Peter Galison, Objectivity. New York: Zone Books, 2007. Print.

Debroise, Olivier. *Mexican Suite: A History of Photography in Mexico.* Trans. and revised Stella de Sá Rego with the author. Austin, TX: U of Texas P, 2001. Print.

Downing, Eric. *After Images: Photography, Archaeology, and Psychoanalysis and the Tradition of Bildung.* Detroit: Wayne State UP, 2006. Print.

Edwards, Gwynne. *Lorca, Buñuel, Dalí: Forbidden Pleasures and Connected Lives.* London and New York: I. B. Tauris, 2009. Print.

"Einstein en los toros." http://www.elsiglodeuropa.es/siglo/historico/Aguilar%202005/651Aguilar.htm. Accessed August 2013. Web.

Elena, Alberto and Javier Ordóñez. "Science, Technology, and the Spanish Colonial Experience in the Nineteenth Century." *Osiris* 2a serie.15 (2000): 70–82. Print.

Everdell, William R. *The First Moderns: Profiles in the Origins of Twentieth-Century Thought.* Chicago and London: U of Chicago P, 1997. Print.

Fontanella, Lee. "Landscape in the Photography of Spain." *Visualizing Spanish Modernity.* Ed. Susan Larson and Eva Woods. Oxford and New York: Berg, 2005. 163–177. Print.

Foster, Hal. *The Anti-Aesthetic: Essays on Postmodern Culture.* Seattle, WA: Bay Press, 1983. Print.

Freeland, Cynthia. *Portraits and Persons: A Philosophical Inquiry.* London: Oxford UP, 2010. Print.

Freire, Miguel, and Pablo García-López, "Cajal: ciencia, arte, ténica de lo microscópico." http://www.scribd.com/doc/33790831/Cajalscienceart microscopictechnique. Accessed July 2013. Web.

Frost, Daniel. *Cultivating Madrid: Public Space and Middle-Class Culture in the Spanish Capital, 1833–1890*. Lewisburg, PA: Bucknell UP, 2008. Print.

Galison, Peter L., and D. Graham Burnett, "Einstein, Poincaré and Modernity: A Conversation." *Daedalus*. Spring 2003. 1–15. Print.

Garagorri, Paulino. "Nota preliminar." José Ortega y Gasset. *Notas de andar y ver: viajes, gentes y paisajes*. Madrid: Revista de Occidente en Alianza Editorial, 1988. 9–12. Print.

García-Ballesteros, Aurora. "Manuel de Terán-Alvarez, 1904–1984." *Geographers. Bibliographical Studies*. Vol. 11. 1987. 145–53. Print.

García López, Pablo, Virginia García Marín, and Miguel Freire. "The Histological Slides and Drawings of Cajal." *Frontiers in Neuroanatomy* 4 (March 2010): 1–16. http://www.ncbi.nlm.nih.gov/pmc/articles/PMC2845060/. Accessed August 2013. Web.

Gilloch, Graeme. *Myth and Metropolis: Walter Benjamin and the City*. Malden, MA: Polity Press/Blackwell, 1996. Print.

———. *Walter Benjamin: Critical Constellations*. Cambridge: Polity, 2002.

Glick, Thomas F. *Einstein in Spain: Relativity and the Recovery of Science*. Princeton, NJ: Princeton UP, 1988. Print.

———. *Einstein y los españoles: ciencia y sociedad en la España de entreguerras*. Madrid: Consejo Superior de Investigaciones Científicas, 2005. Print.

Gómez de la Serna, Ramón. *Movieland*. Trans. Angel Flores. New York: The Macaulay Company, 1930. Print.

Gómez Mendoza, Josefina. "Introducción." Manuel de Terán. *Ciudades Españolas (Estudios de Geografía Urbana)*. Madrid: Real Academia de la Historia. 2004. 11–26. Print.

Goudie, Andrew, and Heather Viles. *Landscapes and Geomorphology: A Very Short Introduction*. Oxford and New York: Oxford UP, 2010. Print.

Grant, Gunnar. "How Golgi Shared the 1906 Nobel Prize in Physiology or Medicine with Cajal." http://nobelprize,org/nobel_prizes/medicine/articles/grant/index.html. 1–3. Accessed August 2013. Web.

Guerra de la Vega, Ramón. *Madrid, años 20: La época de Dalí, Lorca y Buñuel. Historia de la fotografía*. Madrid: Street Art Collection, 2008. Print.

Habermas, Jürgen. "Modernity: An Unfinished Project." *Habermas and the Unfinished Project of Modernity*. Ed. M. P. d'Entréves and S. Benhabib. Cambridge: MIT Press, 1997. 38–55. Print.

Hahn, Otto. "Entrevista a Marcel Duchamp." *Vuelta*. 133–34 (1988). http://ddooss.org/articulos/entrevistas/marcel_duchamp.htm. Accessed July 2013. Web.

Hand, Stacy. "Microphotography." *Encyclopedia of Nineteenth-CenturyPhotography.* Ed. John Hannavy. New York: Routledge, 2007. 925–28. Print.

Hartoonian, Gevork. "Looking Backward, Looking Forward: Delightful Delays." *Walter Benjamin and Architecture.* Ed. Gevork Hartoonian. London and New York: Routledge, 2010. 23–38. Print.

Hempel, Carl Gustav. *Aspects of Scientific Explanation and Other Essays in the Philosophy of Science.* New York: Free Press, 1965. Print.

Holton, Gerald. "Henri Poincaré, Marcel Duchamp, and Innovation in Science and Art." *Leonardo* 34.2 (April 2001): 127–34. March 13, 2006. http://www.mitpressjournals.org/doi/abs/10.1162/002409401750184681?journalCode=leon. Accessed August 2013. Web.

Ibáñez, Juan José. "Suelo y paisaje: una introducción." http://www.madrimasd.org/blogs/universo/2006/06/02/27572. Accessed July 2013. Web.

Interview with Juan A. de Carlos. http://www.desdeelexilio.com?2011/01/16/el-legado-de-cajal-entrevista. Accessed August 2013. Web.

Jay, Martin. *Downcast Eyes: The Denigration of Vision in Twentieth Century French Thought.* Berkeley: U of California P, 1993. Print.

Jones, R. V. "Complementarity as a Way of Life." *Niels Bohr: A Centenary Volume.* Ed. A. P. French and P. J. Kennedy. Cambridge, MA: Harvard UP, 1985. Print.

Keller, Cory. "Sight Unseen: Picturing the Invisible." *Brought to Light: Photography And the Invisible, 1840–1900.* Ed. Cory Keller. New Haven and London: San Francisco Museum of Modern Art and Yale University Press, 2008. 19–35. Print.

Kingsley, Hope. "Art Photography and Aesthetics." *Encyclopedia of 19th Century Photography.* Ed. John Hannavy. New York: Routledge, 2007. 74–82. Print.

Kozloff, Max. *The Theatre of the Face: Portrait Photography Since 1900.* London: Phaidon Press, 2007. Print.

Kurtz, Gerardo F. "Las traducciones al castellano del manual de Daguerre y otros textos fotográficos tempranos en España, 1839–1846." *Archivos de la Fotografía.* Vol. II. No. 1. 1996. http://www.terra.es/personal/gfkurtz/DAGprin.html#Principio. Accessed August 2013. Web.

———. "Spanish Photography." http://www.answers.com/topic/spain. Print.

Lafuente, Antonio, and Tiago Saraiva. "The Urban Scale of Science and the Enlargement of Madrid (1851–1936)." *Social Studies of Science* 34.4 (August 2004): 531–569. Print.

Lahiji, Nadir. "Architecture Under the Gaze of Photography. Benjamin's *Actuality* and Consequences." *Walter Benjamin and Architecture.* Ed. Gevork Hartoonian. London and New York: Routledge, 2010. 75–91. Print.

Laín Entralgo, Pedro, and Agustín Albarracín. *Santiago Ramón y Cajal o la pasión de España.* Barcelona: Ed. Labor, SA, 1978. Print.

Lara López, Emilio Luis. "Historia de la fotografía en España, un enfoque desde lo global hasta lo local." http://clio.rediris.es/no30/emiclio.htm. 1–12. Print.

Larson, Susan, and Eva Woods, eds. "Visualizing Spanish Modernity: Introduction." *Visualizing Spanish Modernity.* Oxford and New York: Berg, 2005. 1–23. Print.

Lavédrine, Bertrand, with Jean-Paul Gandolfo, John McElhone, and Sybylle Monod. Trans. John McElhone. *Photographs of the Past: Process and Preservation.* Los Angeles, CA: The Getty Conservation Institute, 2009. Print.

Leach, Andrew. "Manfredo Tafuri and the Age of Historical Representation." *Walter Benjamin and Architecture.* Ed. Gevork Hartoonian. London and New York: Routledge, 2010. 5–21. Print.

Lindley, David. *Uncertainty: Einstein, Heisenberg, Bohr, and the Struggle for the Soul of Science.* New York: Random House, 2008. Print.

Llinás, Rodolfo R. "The Contribution of Santiago Ramón y Cajal to Functional Neuroscience." *Nature* 4 (January 2003): 77–80. www.nature.com/reviews/neuro. Accessed July 2013. Web.

López Mondéjar, Publio. *150 años de fotografía en España.* 3rd ed. Barcelona: Ed. Lunwerg, 1999. Print.

Malvasi, Mark G. "Historical Consciousness and Its Enemies." *The University Bookman* 44. 3 (Summer 2006): 1–8. http://www.kirkcenter.org/index.php/bookman/article/historical-consciousness-and-its-enemies/. Accessed July 2013. Web.

Martineau, Paul. *Still Life in Photography.* Los Angeles: The J. Paul Getty Museum, 2010. Print.

Martínez de Pisón, Eduardo. "La perspectiva naturalista." *Manuel de Terán Geógrafo1904–1984.* Ed. Eduardo Martínez de Pisón and Nicolás Ortega Cantero. Madrid: Residencia de Estudiantes/Sociedad Estatal de Conmemoraciones Culturales, 2007. 111–34. Print.

Martínez de Pisón, Eduardo, and Nicolás Ortega Cantero. "Manuel de Terán, geógrafo (1904–1984): Exposición en la Residencia de Estudiantes (marzo-junio 2007)." *Anales de Geografía* 27.2 (2007): 179–86. Print.

Martínez-Frías, J., D. Hochberg, and F. Rull. "A Review of the Contributions of Albert Einstein to Earth Sciences—In Commemoration of the World Year of Physics." *Naturwissenschaften* 93(2): 66–71. Accessed June 2013. http://www.ncbh.nlm.nih.gov/pubmed/16453104. Web. February 2006.

Mattos, Claudia. "Landscape Painting Between Art and Science." *Alexander von Humboldt: From the Americas to the Cosmos.* Ed. Raymond Erickson, Mauricio A. Font, and Brian Schwartz. New York: Bildner Center for Western Hemisphere Studies, Graduate Center The City U of New York, 2004. 141–55. Print.

Mendelson, Jordana. *Documenting Spain: Artists, Exhibition Culture, and the Modern Nation, 1929–1939*. University Park, PA: The Penn State UP, 2005. Print.

Monsiváis, Carlos. *Quietecito por favor!* Mexico, DF: Grupo Carso, 2005. Print.

Montes-Santiago, J. "El encuentro de Einstein y Cajal (Madrid, 1923): Un olvidado momento estelar de la humanidad." *Revista de Neurología* 43 (2006): 113–17. Print.

Nieto Olarte, Mauricio. "Scientific Instruments, Creole Science, and Natural Order in the New Granada of the Early Nineteenth Century." *Journal of Spanish Cultural Studies* 8.2. (July 2007): 235–252. Print.

Nietzsche, Friedrich. *The Will to Power*. Trans. Walter Kaufmann and R. J. Hollingdale. New York: Vintage, 1967. Print.

Ortega y Gasset, José. "Bronca en la física." *Meditación de la técnica y otros ensayos sobre ciencia y filosofía*. Madrid: Revista de Occidente en Alianza Editorial, 1988. 143–63. Print.

———. "De Madrid a Asturias o los dos paisajes." *Notas de andar y ver. Viajes, gentes y paisajes*. Madrid: Revista de Occidente en Alianza Editorial, 1988. 75–96. Print.

———. "Las ermitas de Córdoba." *Notas de andar y ver. Viajes, gentes y paisajes*. Madrid: Revista de Occidente en Alianza Editorial, 1988. 13–17. Print.

———. "Las fuentecitas de Nuremberga." *Notas de andar y ver. Viajes, gentes y paisajes*. Madrid: Revista de Occidente en Alianza Editorial, 1988. 21–27. Print.

———. "Primera escaramuza con el tema." *Meditación de la técnica y otros ensayos sobre ciencia y filosofía*. Madrid: Revista de Occidente en Alianza Editorial, 2008. 21–29. Print.

———. "¿Qué es la técnica? Introducción: La universidad y la técnica." *Meditación de la técnica y otros ensayos sobre ciencia y filosofía*. Madrid: Revista de Occidente en Alianza Editorial, 2008. 13–19. Print.

———. *La rebelión de las masas*. Madrid: Revista de Occidente en Alianza Editorial, 2006. Print.

———. "Sobre la expresión, fenómeno cósmico." *El Espectador VII. Obras Completas*. Madrid: Fundación José Ortega y Gasset/Taurus, 2004. Tomo II. Print.

———. "Sobre el punto de vista en las artes." http://www.scribd.com/doc/3819932/Ortega-y-Gasset-Jose-Sobre-el-punto-de-vista-en-las-artes. Accessed July 2013. Web.

———. "Temas del Escorial." *Notas de andar y ver. Viajes, gentes y paisajes*. Madrid: Revista de Occidente en Alianza Editorial, 1988. 43–74. Print.

———. "El tecnicismo moderno—los relojes de Carlos V—ciencia y taller—el prodigio del presente." *Meditación de la técnica y otros ensayos sobre ciencia y filosofía.* Madrid: Revista de Occidente en Alianza Editorial, (1933) 2008. 91–96. Print.

———. "Tierras de Castilla. Notas de andar y ver." *Notas de andar y ver. Viajes, gentes y paisajes.* Madrid: Revista de Occidente en Alianza Editorial, 1988. 34–42. Print.

———. "'Viaje de España.'" *Notas de andar y ver. Viajes, gentes y paisajes.* Madrid: Revista de Occidente en Alianza Editorial, 1988. 28–33. Print.

———. "Vicisitudes en las ciencias." *Meditación de la técnica y otros ensayos sobre ciencia y filosofía.* Madrid: Revista de Occidente en Alianza Editorial, (1930) 2008. 135–41. Print.

Otis, Laura. *Membranes: Metaphors of Invasion in Nineteenth-Century Literature, Science, and Politics.* Baltimore and London: Johns Hopkins UP, 1999. Print.

Paredes Martín, María del Carmen. "Elementos para una teoría del paisaje en Ortega y Gasset." *El primado de la vida: cultura, estética y política en Ortega y Gasset.* Ed. Atilano Domínguez, Jacobo Muñoz, and Jaime de Salas. Cuenca: Ediciones de la Universidad de Castilla–La Mancha, 1997. 177–93. Print.

Parkinson, Gavin. *Surrealism and Modern Science: Relativity, Quantum Mechanics, Epistemology.* New Haven and London: Yale UP, 2008.

Phillips, Christopher. "The Judgment Seat of Photography." *The Contest of Meaning: Critical Histories of Photography,* ed. Richard Bolton. Cambridge, MA: MIT Press, 1989. Print.

Potolsky, Matthew. *Mimesis (The New Critical Idiom).* New York and London: Routledge, 2006. Print.

Portela, Eugenio and Amparo Soler, "La química española del siglo XIX." *Ayer* 7 (1992): 85–107. Print.

Pratt, Dale J. *Signs of Science: Literature, Science, and Spanish Modernity Since 1868.* West Lafayette, IN: Purdue UP, 2001. Print.

Prodger, Phillip. *Darwin's Camera: Art and Photography in the Theory of Evolution.* Oxford: Oxford UP, 2009. Print.

Ramón y Cajal, Santiago. "La fotografía." *El mundo visto a los ochenta años: impresiones de un arteriosclerótico.* Madrid: MAXTOR, 2008. Tercera edición. Facsimile of 1939 edition. 157–59. Print.

———. *Fotografía de los colores: bases científicas y reglas prácticas.* Zaragoza: Prames—Las tres sorores y Gobierno de Aragón, 1966. Facsimile of 1912 ed. Print.

———. *Mi infancia y juventud. Ramón y Cajal: Obras literarias completas.* Madrid:Aguilar, 1947. 25–275. Print.

———. *El mundo visto a los ochenta años: impresiones de un arteriosclerótico.* Tercera edición. Madrid: MAXTOR, 1939 (facsimile of 1939 edition). Print.

————. *La psicología de los artistas.* 2nd ed. Buenos Aires: Espasa Calpe Argentina, SA, 1954. Print.

————. "Prólogo. Madrid 22 de abril 1904." Intro. José Puche Alvarez, "Presentación de un ensayo poco conocido de Cajal." *Vivencias de Don Santiago Ramón y Cajal.* Ed. Manuel Peláez Cebrián. Mexico: Sociedad Médica Hispano Mexicana, Universidad Nacional Autónoma de México, Instituto Politécnico Nacional, Fondo de Cultura Económica, 1999. 75–86. Print.

————. *Recuerdos de mi vida.* Partes I, II, III. http://cvc.cervantes.es/ciencia/cajal/cajal_recuerdos/recuerdos/labor_03.htm. Accessed August 2013. Web.

————. *Reglas y consejos sobre la investigación científica: los tónicos de la voluntad.* Prol. Severo Ochoa. Madrid: Espasa Calpe, 1941. 21st ed. 2008. Print.

Ramón y Cajal Junquera, Santiago. "Ramón y Cajal, microbiologist." *International Microbiologist* 3 (2000) 59–61. http://revistes.iec.cat/index.php/IM/article/view/4c457c0c4bbe7.002/9242. Accessed August 2013. Web.

Romero, Alfredo. Introducción, catalogación y notas. *Fotografía Aragonesa I, Ramón y Cajal.* Zaragoza: Diputación Provincial de Zaragoza, 1984. 1–104. Print.

Ross, Christopher J. *Spain, 1812–1996: Modern History for Modern Languages.* London: Arnold, 2000. Print.

Rossi Monti, Martino. "Snowflakes and Spiritual Exercises." Martino Rossi Monti, Patrick Singy, and Albena Yaneva, "On Lorraine Daston and Peter Galison's *Objectivity*, MIT Press, 2007." *Iris* I (April 2009): 277–88. Print.

Ruffa, Astrid. "Dalí's Surrealist Activities and the Model of Scientific Experimentation." *Papers of Surrealism* 4 (Winter 2005): 1–14. http://www.surrealismcentre.ac.uk/papersofsurrealism/journal4/acrobat%20files/ruffapdf.pdf. Accessed August 2013. Web.

Ruiz, Carme. "Salvador Dalí and Science." *El Punt* October 18, 2000. http://www.salvador-dali.org/serveis/ced/articles/en_article3.html. Accessed August 2013. Web.

Saldaña, Diana. "Spain." *Encyclopedia of Nineteenth-Century Photography.* Ed. John Hannavy. New York: Routledge, 2007. 1322–9. Print.

Schwartz, Egon. "Ortega y Gasset and German Culture." *Monatshefte* 49.2. (February 1957): 87–91. Print.

Sontag, Susan. "Introduction." Walter Benjamin, *One-Way Street and Other Writings.* Trans. Edmund Jephcott and Kinglsey Shorter. London: Verso, 1985. 7–28. Print.

Strong, Tom. "A Review of Lorraine Daston and Peter Galison's *Objectivity.*" *The Weekly Quantitative Report* 1.10 (December 2008): 62–66. Print.

Terán, Manuel de. *Ciudades Españolas (Estudios de Geografía Urbana)*. Madrid: Real Academia de la Historia, 2004. Print.

———. "Prólogo." Julio Vinuesa Angulo, *El desarrollo metropolitan de Madrid: Sus repercusiones geodemográficas*. Madrid: Instituto de Estudios Madrileños, 1976. 9–12. Print.

———. "Una ética de conservación y protección de la naturaleza." *Manuel de Terán Geógráfo 1904–1984*. Eds. Eduardo Martínez de Pisón and Nicolás Ortega Cantero. Madrid: Residencia de Estudiantes/Sociedad Estatal de Conmemoraciones Culturales, 2007. 377–91. Print.

Triarhou, Lazaros C., and Ana B. Vivas. Introduction and translation, "Poetry and the Brain: Cajal's Conjectures on the Psychology of Writers." *Perspectives in Biology and Medicine* 52.1 (Winter 2009): 80–89. Print.

Ugarte, Michael. *Madrid 1900: The Capital as Cradle of Literature and Culture*. University Park, PA: Pennsylvania State UP, 1996. Print.

Vizuete, Pelayo. *Einstein y el misterio de los mundos*. Madrid: Editorial Arte y Ciencia, 1923. Print.

Weidemann, Christiane. *Salvador Dalí*. New York: Prestel, 2007. Print.

Index